Cost Engineering

Cost Engineering

A Practical Method for Sustainable Profit Generation in Manufacturing

Chris Domanski

CRC Press
Taylor & Francis Group
Boca Raton London New York

CRC Press is an imprint of the
Taylor & Francis Group, an **informa** business

CRC Press
Taylor & Francis Group
6000 Broken Sound Parkway NW, Suite 300
Boca Raton, FL 33487-2742

First issued in paperback 2021

© 2020 by Taylor & Francis Group, LLC
CRC Press is an imprint of Taylor & Francis Group, an Informa business

No claim to original U.S. Government works

ISBN 13: 978-1-03-224330-6 (pbk)
ISBN 13: 978-0-367-44083-1 (hbk)

Visit the Taylor & Francis Web site at
http://www.taylorandfrancis.com

and the CRC Press Web site at
http://www.crcpress.com

Thank you to my incredible wife Angie for our beautiful kids, Sophia and Evan, and for allowing me to take time away from our family so that I could write this book.

Thank you to Douglas T. Hicks for further igniting my curiosity and passion for cost engineering, and ultimately inspiring me to write this book.

Thank you again to Douglas T. Hicks, as well as Leonard Magro, and Nazire Yaman for their feedback and guidance.

Contents

Preface

According to the Small Business Administration (SBA) Office of Advocacy's 2018 Frequently Asked Questions, only about a half of small businesses survive past the first 5 years and only about a third past the first 10 years. According to trendingeconomics.com, the United States alone averaged 44,500 business bankruptcies per year from 1980 to 2019. That equates to over 2.6 million bankruptcies. The results have been devastating to millions of people worldwide. Trillions of dollars have been lost by individuals and stakeholders.

There are many reasons for all these failures, but one that is not mentioned often enough is the lack of cost awareness and cost management. I have not been able to find any studies that try to quantify this effect, but based on my 25-plus years of experience in the manufacturing industry and dealings with a multitude of manufacturing companies, I am certain that hundreds of billions of dollars are lost every year due to poor cost management. I would estimate that at least three quarters of manufacturing companies do not fully understand their own costs, and make decisions daily that negatively impact their profitability. In addition, I estimate that roughly 90 percent of manufacturing companies do not have an existing or effective cost engineering process to assure sustainable profitability over time.

What is even more staggering is that there are almost zero publicly available publications or training to help these manufacturing companies. Whatever available knowledge and training, which is very limited, is contained within each organization and the individuals that perform cost management and cost engineering jobs. Mastery is acquired in real-time, over the course of many years, with individuals and departments stumbling through raw data that eventually adds up to something akin to "tribal knowledge." This knowledge may or may not reflect an accurate way to cost manage and cost engineer products. Usually, it is not. Or, it is vastly incomplete.

This is why I decided to write this book. I wanted to help companies and people working for them – from engineers, buyers, accountants, cost estimators, and their managers, all the way up to executives – to quickly advance their cost management and engineering knowledge. I also wanted to show them the right way of doing things while at the same time making them aware of the wrong ways. Ultimately, my objective is to improve the profitability of manufacturing industries so that they have greater chance of maintaining and creating jobs. I have witnessed too many times, incompetent or simply unaware individuals who have taken their companies down the path of profit destruction that eventually led to people losing their jobs. This can all be avoided or at least minimized with better knowledge. This book will hopefully empower individuals and corporations to better their understanding of costs, thereby reducing the chance of business failure.

About the Author

Mr. Chris Domanski has over 25 years of experience in Cost Engineering, Purchasing, Finance, and Engineering in the automotive industry, most recently as a Senior Manager of Purchasing and Cost Optimization at Nexteer Automotive. Mr. Domanski has had the fortune of working for many great companies, large and small, such as Ford Motor Company, Continental Automotive, TRW, ZF Group, Methode Electronics, and Nexteer Automotive. Mr. Domanski had an opportunity to work for both customers and suppliers within the industry, thus sat on both sides of the negotiation table. In his career, he has been responsible for quoting, where he developed business cases and cost allocation models; cost estimating, where he analyzed over 300 supplier-manufacturing plants and their cost structures; and cost optimization, where he was responsible for finding and implementing cost reduction initiatives. This unique experience has given Mr. Domanski an unmatched level of expertise in all areas of the cost engineering discipline. He is thankful that he has been able to help many companies become more profitable and thus save and create a lot of jobs. He hopes that he also inspired others to do the same. To spread awareness of cost engineering principles and thus improve efficiency of all manufacturing industries, Mr. Domanski also administers a LinkedIn group "Manufacturing Cost/Price Estimators, Engineers, Analysts, and Controllers" with over 5,200 members, and is a frequent speaker at the annual Automotive Cost Engineering Conference in Detroit, MI.

Mr. Domanski graduated with a Bachelor of Science Degree in Mechanical Engineering from Wayne State University in Detroit, MI, and a Master's Degree in Business Administration with Finance concentration from Oakland University in Rochester Hills, MI. He also holds a certificate in Negotiation and Leadership from Harvard University Law School in Boston, MA.

1 What Is Cost Engineering?

Cost engineering is defined by Wikipedia as "the engineering practice devoted to the management of project cost, involving such activities as estimating, cost control, cost forecasting, investment appraisal, and risk analysis." In simpler terms, cost engineering (sometimes called design to cost) is the practice of engineering a company's products to meet pre-defined cost requirements. It is really an umbrella of various methodologies (see Figure 1.1) that are often confused for cost engineering itself, but really only address portions of it that have something to do with cost estimating, cost control, or cost optimization.

Companies often believe they have cost engineering capabilities when in fact they actually have none or, at best, a subset of those capabilities. In the following sections we will discuss some of those capabilities including target costing, cost estimating, should costing, cost modeling, activity-based costing, value engineering and value analysis (VA/VE), theory of the resolution of invention related tasks (TRIZ), cost and value management, cost accounting, standard costing, and marginal costing.

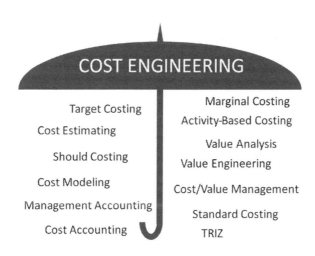

FIGURE 1.1 Umbrella of cost engineering methodologies.

TARGET COSTING

This term refers to the cost controlling practice of setting cost targets for every component of a product (e.g., parts that go into a vehicle) during the product development process. Although this concept was first developed in the United States, it was the Japanese car manufacturers that made it a very regimented and disciplined practice. It is a methodology quite similar to cost engineering, and some would argue that target costing and cost engineering are one and the same, because target costing is usually tied with the Japanese kaizen, or continuous improvement, methodology. This means that not only are cost targets set, but also kaizen is used to meet or improve on those targets. Cost engineering, however, is more encompassing and embraces virtually all methodologies that revolve around engineering a product to a specific cost target.

There are several good books on target costing that give detailed explanation of the process, such as "Target Costing and Kaizen Costing" by Yasuhiro Monden and "Target Costing: The Next Frontier in Strategic Cost Management" by Shahid L. Ansari and Jan E. Bell.

COST ESTIMATING

The concept of cost estimating refers specifically to estimating the cost of products. The ability to accurately estimate the cost of products – both those being purchased and those being produced – can and should be a critical element in the process of cost engineering. It is not, however, the only thing required in cost engineering, because estimating does not encompass cost optimization concepts.

It is important to note that some companies refer to the practice of referencing historical costs as the practice of cost estimating. That could be useful information, but what is often needed, and what is often more accurate, is estimating the cost from the "ground up," meaning estimating each cost element separately until it builds up to a price.

For an automotive company, the ability to accurately estimate each supplier's cost of producing its components, as well as the cost for the company to assemble the vehicle, would be critical knowledge in developing targets. In fact, due to the critical nature of cost estimating in product development and the complexity and skill required to achieve accuracy in cost estimating, several chapters will be devoted to this topic.

SHOULD COSTING

The term "should costing" is often used interchangeably with cost estimating, and they are one and the same. However, using the word "should" is meant to imply that whatever is estimated is accurate and will be the actual cost. Since some people have a negative association with the word "estimate," associating it with guessing, the use of "should" seems to have been intentionally used to infer accuracy and attainability. However, just like cost estimating, should costing simply uses assumptions to arrive at an estimate for what a product should cost, and both have the same accuracy limitations.

COST MODELING

The term "cost modeling" is often used interchangeably with cost estimating; however, it often is a more in-depth analysis of the company's or its suppliers' costs. This will be discussed in more detail in Chapter 2, but cost estimating might require some assumptions due to the unavailability of detailed information, whereas cost modeling typically will try to model exactly where the costs should be assigned based on historical financial and operational information (profit and loss or P&L statement). Although some assumptions will still be made, as long as the right activities are defined and cost drivers used, the cost model methodology should lead to better accuracy and predictability since the model will represent economic reality for any given company. On the other hand, if the data are flawed, the wrong activities are defined, or the wrong cost drivers are used, this methodology will lead to very incorrect results.

Another important differentiator of cost modeling is that it tries to predict the impact of changes to the cost structure. Whereas cost estimating typically uses the cost information available at that point in time, the cost model will adjust the cost structure based on the changes in business such as changes in capacity utilization or scale of manufacturing operation. For example, if a business being quoted improves the factory utilization from 65% to 80%, the improved absorption of fixed costs will be accounted for in a cost model, whereas a cost estimate might still assume 65% capacity utilization because that is the status quo at that particular time of estimation.

ACTIVITY-BASED COSTING (ABC)

Perhaps the most misunderstood term among many cost engineering–related terms, activity-based costing (ABC) refers to a specific method of estimating or modeling cost. Although cost estimating already centers around costing activities, the ABC methodology describes a more formal and focused way of understanding how the activities of an organization relate to its products or services and then assigns all costs that a company incurs to those activities based on causality and from those activities to the products or services they support. It is similar to cost modeling, but has fewer assumptions and more cost buckets distributed using cost drivers.

For example, in a car manufacturing company, one obvious activity is the assembly of the vehicle itself. It is obvious because it is easily visible to all those who work for the company, since each car must be assembled before it can be shipped off for sale. Very often, the manufacturing process is where companies end their application of ABC; however, there are many activities that happen before and after the car is assembled, such as design development, product launch, parts procurement, accounting of all financial transactions, sales and marketing, production scheduling, and many more. These are not directly related to assembling the vehicle, but they are activities that cost money.

If a company makes only one product, then all those costs can simply be allocated to that one product. However, what if a car company decides to make trucks in addition to cars? The cost for these activities will be different for the two different vehicles (e.g., marketing of one will probably be more or less expensive than the other), so the question becomes how to assign costs for various activities to the two different vehicles, thereby making sure the profitability of each is well understood and that the right financial decisions are made for the company.

This is where ABC is most useful, because it helps to build cost models that accurately predict the flow of costs and ultimately product profitability in fairly complex situations. There are many books written about ABC, but Douglas T. Hicks' "Activity-Based Costing: Making It Work for Small and Mid-Sized Companies" is recommended for its practical, hands-on explanation.

VALUE ANALYSIS AND VALUE ENGINEERING (VA/VE)

VA/VE is an advanced methodology of product cost optimization that has been deployed in many industries. Its focus is on improving value to the customer with value being defined as function divided by cost, meaning that increasing product function and decreasing product cost improve value.

The difference between value engineering and value analysis is that the first deals with value improvement during the product development process while the latter occurs after the product is launched. In both value engineering and value analysis, the process of optimizing a product is virtually the same. It starts with a cost reduction idea generation where a person or a team of people evaluates designs that are being developed or in production. There are many ways to spark idea generation, such as TRIZ or benchmarking (both explained later); however, the VA/VE purists primarily use functional analysis for this purpose. This method is used to identify every function that a product performs, assign cost to each function, and then brainstorm to find ways to perform the same or better function at a lower cost.

For example, the main function of a car engine is to propel a vehicle down the road. A hybrid car might do this with both a gasoline engine and battery-powered motor(s). However, what if those could be replaced with a new technology using four smaller motors/battery packs, one at each wheel, and eliminate the gasoline engine altogether? The same function could be performed, perhaps with increased functionality (quicker acceleration) and decreased cost (no gasoline engine), thus increasing the value.

There are many books on VA/VE methodology, starting with "Techniques of Value Analysis and Engineering" by Lawrence D. Miles, who is considered the founder of this methodology, and also a professional organization called SAVE International that fosters these methods in many industries such as civil engineering, aerospace, defense, automotive, and others. Although VA/VE is a stand-alone methodology, its concepts are a vital part of the overall cost engineering field, which later chapters will explain in detail.

TRIZ

Since TRIZ (from Russian for theory of the resolution of invention related tasks) has gained popularity in some industries, it is worth explaining its meaning. Developed by a Russian inventor, Genrich Altshuller, during the Soviet Union period, this problem solving methodology is often used to brainstorm for cost optimized design or manufacturing process solutions. What Altshuller discovered, based on his analysis of historical patents, is that problems almost always were solved one of 40 different ways, or principles. His methodology was to apply these principles on any new problem until a solution was found.

For example, one typical solution is merging, meaning that perhaps a solution to a problem is to merge several systems together into one to generate synergies. More specifically, a solution to a data computing limit might be to combine a computing power of several computers. Or, a single processor chip might be used to perform many different functions.

There are several books on TRIZ, such as "And Suddenly the Inventor Appeared" by Altshuller himself. There are even online apps that help guide an inventor or a problem solver through the TRIZ brainstorming process.

COST AND VALUE MANAGEMENT

The terms "cost management" and "value management" are very broad and refer to any of the company's activities related to improving cost and/or value. They could mean employing any of the previously described techniques or none of them. Sometimes companies refer to cost management as nothing more than tracking their costs to make sure budgets are adhered to. Or, with value management, it could simply mean that the company is stream mapping the value chain, meaning that they are minimizing the number of steps in producing a product from raw material to finished product.

MANAGEMENT ACCOUNTING

"Management accounting" is a very broad term that refers to using accounting information to provide management with decision-making capability. Instead of a purely historical view of accounting where only past spending is categorized in GAAP (generally accepted accounting principles)-defined buckets, the goal of management accounting is to also provide more insight into where the money was and will be spent and where it might make sense to make adjustments in the future.

Management accountants and their primary organization, called the Institute of Management Accountants (IMA), consist mostly of accountants working in the accounting industry and try to raise awareness of the need for more management-friendly accounting information. Their suggested main tool for achieving this is ABC, which is part of the cost engineering tool set.

COST ACCOUNTING

As the name implies, cost accounting is an accounting practice that involves setting up a system to track costs as they occur. However, this practice is more involved than simply managing a company's ledger, since it tries to track costs more accurately and tie them to certain areas or departments within a company. Although some would argue that this tool helps reduce cost because it gives better visibility of where costs occur, this is not necessarily a cost engineering tool, because it may not be required to design products to specific cost targets. It is more of an accounting tool that helps accountants monitor and address spending.

STANDARD COSTING

Standard costing is a cost accounting tool that helps a company track costs. The difference here to cost accounting is that standard costing defines budgeted costs for every product and manufacturing step on an hourly basis, then it tracks and compares actual costs versus those budgeted, or standard, costs. For example, a company might set a standard to assemble a vehicle in one hour with a specific cost associated to that one standard hour. If a vehicle is assembled in longer than one hour, then the company knows that it has gone over budget and will need to find a corrective action to get back to budget.

Although this accounting method might be useful in tracking costs and monitoring budgets, it is not necessarily a tool that needs to be used in the cost engineering process. Similar to cost accounting, this is more of an accounting tool.

MARGINAL COSTING

Marginal costing is a cost accounting method based on a concept that only costs that vary with any particular decision should be included in the decision analysis. For example, fixed costs are not relevant to a decision for cases that involve relatively small variations from existing practice and/or are for relatively limited periods of time. For a company producing 300,000 units that require $1,000,000 in variable cost and $2,000,000 in fixed cost to produce, adding another 10,000 units of production will only increase the variable cost by about $50,000, while the fixed cost remains unchanged, resulting in a cost of $5 per unit ($50,000/10,000 units) for those additional 10,000 units. If a typical full absorption method was used to assign unit cost, then the cost per unit would be $10 ($3,000,000 total cost/300,000 units), which would overestimate the unit cost because it would include fixed cost impact that is not real.

This method is sometimes useful for managerial decision-making because it provides a clearer picture of what is the real incremental cost impact of a decision. However, for cases where there are significant changes to assumptions, this cost accounting method can be inaccurate and will model the cost incorrectly.

SPARK EXAMPLE

Despite its seemingly straightforward objective, cost engineering is very difficult to implement successfully. The first challenge is that companies struggle even with the basic task of defining what the cost requirements should be. Companies often know roughly what the market price is of their products because they know at which price points their products will or will not sell in the open market. They also know what profit margins they want or need to make. Market price minus the desired profit margin becomes the cost a product cannot exceed in order for the company to earn its target profit. This is called its "target cost." Pretty straightforward so far, at least for a company that is selling a product with just one item in its bill of materials (BOM) like a plastic toy duck. But what if the product is more complex like a car? What should be the cost of each of the 10,000 individual components that assemble into a car, such as a steering wheel, tires, windows, etc.?

This is where things get complicated. A company that has been in business for some time will have historical cost data available that can be used as a starting target. But what if it is a new company launching a new product or an existing company launching a new or significantly different design? For example, a fictional startup company called SPARK is trying to launch a hybrid truck with both a gasoline engine and an electric motor as powertrains. The SPARK executives think that they can sell their full capacity of 10,000 hybrid trucks per year for $100,000 each. The company's owners need to make a 10% profit in order to get the desired payback on their investment, which means that the cost per vehicle cannot exceed $90,000.

Based on this calculation, the company knows that they have a $90,000 target cost for their vehicle. After its engineering team spends about two years and $20,000,000 (100 people × $100,000 average salary × 2 years) developing the vehicle, the company has a preliminary design that meets both its design and performance requirements. However, after the Purchasing department sends out requests for quote (RFQs) to various component suppliers and spends about three months collecting data, the total vehicle BOM cost adds up to $115,000. In addition, the company estimates it will spend about $15,000 per vehicle in manufacturing and corporate costs ($150,000,000 divided by 10,000 vehicles annually). This results in a total estimated vehicle cost of $130,000, or $40,000 over the target cost. What should the company do now?

SPARK's finance team goes back to Engineering and challenges them to reduce costs, but each component engineering group refuses to change the design. To do so, they all claim, would make it impossible for them to meet the vehicle's performance requirements. The gasoline powertrain engineers must have certain power output, so only a four-cylinder engine with 250 hp will do. The electric motor powertrain engineers must also meet their power output requirements, so they cannot compromise on the battery size. The same feedback comes from chassis, body, brakes, and other engineering teams. Since redesign doesn't appear to be a viable option, the finance team goes back to Purchasing and challenges them to negotiate better pricing with suppliers. Surely, Engineering could not have designed this truck so badly. This must be Purchasing's fault. The Purchasing team then goes back to suppliers and uses every negotiation tactic they know to get suppliers to lower their prices. Since SPARK has not produced a single vehicle yet and 10,000 parts is not a lot of business for most suppliers, the Purchasing team lacks leverage and is only able to lower the BOM cost down to $110,000. The vehicle is still $35,000 over the target.

Unfortunately, this is a very common way for things to unfold. The result is often an unsuccessful product launch and, in the worst case, the company's bankruptcy. We will later go through the details of an effective way of developing products to target, but in simple terms the SPARK engineering team should have worked together up front with other company functions to design each of the vehicle components with specific performance and cost targets in mind.

SUMMARY

Cost engineering is focused on developing products to specific cost requirements and is an umbrella of concepts relating to cost estimating, control, and optimization. These concepts include, but are not limited to, Target Costing, Cost Estimating, Should Costing, Cost Modeling, Activity Based Costing, Value Analysis, Value Engineering, TRIZ, Cost and Value Management, Cost Accounting, Standard Costing, and Marginal Costing.

Although cost engineering is a simple concept, its implementation is complex and must be carefully orchestrated in order to be successful. It involves almost all functions of the organization such as sales, finance, engineering, operations, and purchasing. It also touches every part of the product development process, from concept development to product manufacturing.

It is important for a company to start the cost engineering process early in the product development. Waiting to cost optimize products until after design is finalized could be a fatal mistake. There will not be enough time and resources to make any significant changes and the price that a product can bear in the market will not change, so the result will be diminished profit or possibly financial loss. With early focus on cost engineering, the development team can be made aware of cost issues and have time to adjust design direction. This will result in a product that meets its cost and profit targets while maintaining an acceptable market price.

2 Cost Estimating Basics

The capability to accurately estimate cost is the most important factor in the cost engineering process. If a company implements a cost engineering methodology, it will already be on its way to higher profitability, but it will not be able to make the right decisions without accurate cost estimates. Just as a navigator needs navigational tools to guide him to his desired destination, so does the product development team needs cost estimates to guide it in developing profitable products.

This is one of several chapters devoted to cost estimating. Although a great deal of detail will be covered here, it is worth noting that a whole book would be needed to cover this topic in its entirety. Cost estimators spend years mastering the many intricacies of various manufacturing processes such as machining, casting, molding, stamping, and many others. Let us start with the basics.

BASIC COST CATEGORIES

There are various costs that are required to manufacture parts, and not all companies use the same categories or "buckets" to define them all. For the purpose of this book, the categories that are most commonly used and necessary in cost estimating are described below. Figure 2.1 also shows the basic flow of those categories in sequence that the product sees during manufacturing. Each of these categories will be described in detail.

- Raw material
- Purchased components
- Packaging (inbound)
- Logistics (inbound)
- Material overhead
- Manufacturing labor
- Manufacturing overhead
- Setup
- Equipment depreciation
- Scrap
- Corporate overhead
- Profit
- Packaging (outbound)
- Logistics (outbound)
- Tooling and fixtures (not in piece price)

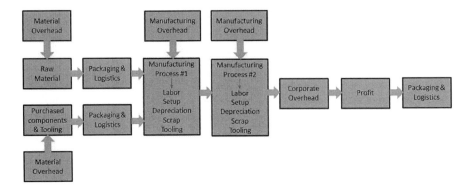

FIGURE 2.1 Cost categories for manufactured products.

RAW MATERIALS

This is the price paid for the basic materials, in their pure form, needed to produce a part. These could be steel, copper, resin, aluminum, or other basic raw materials. Most of these materials are not actually in their purest raw form (e.g., copper dug up directly from the mine), but instead they are received in some pre-manufactured form. For example, a stamping company will most likely receive from its suppliers a type of copper that is alloyed with some other minerals and that is pre-formed into sheets or coils. This means that their supplier or sub-suppliers downstream in the supply chain melted the raw mined mineral, mixed it with some scrap and other materials, and then formed them into shape by melting and pouring it into form.

The stamping company, however, would consider coiled steel its raw material. Similarly, for a molder of plastic components, resin would be considered a raw material, even though this resin is already pre-formed into pellets that were produced by the supplier from oil, glass fibers, and other basic raw materials.

It is important to note that there is raw material scrap, called offal, associated with most manufacturing processes. This is material that is discarded during manufacturing because of the mismatch of incoming and outgoing shapes for the products. In the stamping company example, the copper is received in coils but the parts that are stamped out could be of various shapes and sizes. Nevertheless, the cost of gross weight should be assumed, not the cost of final net weight (weight of final part). This will be further explained in another chapter of this book.

PURCHASED COMPONENTS

These are purchased items that have been pre-made into component parts prior to receipt by the final assembly factory. They are made by sub-suppliers and could include things like screws, bolts, connectors, or others. In the case of a car manufacturer, purchased components are often sub-assemblies that go into a vehicle, such as steering wheel assemblies, radios, HVAC controller assemblies, seat assemblies, or even engines.

PACKAGING (INBOUND)

This is the price paid for packaging materials used for transportation of raw materials and purchased components. This could include cardboard boxes, plastic trays, shrink wrap, wooden pallets, plastic bags, and various other options. What type of packaging is chosen depends on the volume of parts ordered and the distance the parts will travel. If many parts will be ordered over a long period of time, usually the *returnable packaging* option will be chosen. This means that plastic trays will be molded to carry parts and these trays will be returned to the supplier and reused to carry more parts. Even though there is an investment in a tool to make the trays, it has a good return on investment (ROI) since the same trays can be reused multiple times making the packaging cost per part cheaper over time.

On the other hand, if only few parts will be ordered or parts will be traveling overseas, making returning of trays less practical, then most likely cardboard packaging will be used. This is often called *expendable packaging* because it is either thrown away after use or sent to paper recycling for a small credit.

LOGISTICS (INBOUND)

This is the price paid for the freight and duties (tariffs) required in transporting incoming raw materials and purchased components. There are various incoterm (international commercial term) options but in the case where suppliers are responsible to bring their material to their customer, they will include this charge in their cost. Otherwise, the customer will be responsible for the cost.

The incoterm choice depends on many factors. Sometimes it is based on cost, but often it is based on desired inventory levels or risk associated with inventory carrying and transportation. If the choice is purely based on cost, the company with the larger volumes of product will usually have better leverage on logistics cost. A customer might choose a supplier to pay for the logistics if the supplier actually transports more product than the customer. In cases where inventory is the main consideration, the customer might require the supplier to take care of logistics because in those cases the parts are not considered as part of the customer inventory cost and risk until they arrive at the customer factory. The International Chamber of Commerce (ICC) website contains a wall chart of commonly used incoterms with descriptions of carrier and insurance responsibilities.

It is also important to note that customs duties can be a significant cost driver. There are many different tariffs that can be applied to products crossing international borders. For example, during President Donald Trump's administration, tariffs were imposed on many products imported from China. These could be up to 25% on product price, so a part that is priced at $10 suddenly increased in cost to $12.50. This makes it important to account for tariff cost when deciding on purchased materials sources.

In addition to duties, there are also many costs associated with maintaining an overseas supply chain, such as the cost of breaking down and sorting items received in bulk, the cost of maintaining a foreign representative, the cost of sending representatives to other countries for inspections and audits, and the general cost inefficiencies while working with different cultures and time zones. As long as these costs are considered, they should be captured in the logistics or material burden categories.

MATERIAL OVERHEAD

This is the cost of warehousing material and handling it within a manufacturing plant prior to its introduction into production. Although there are differences in accounting practices in various countries, anything that has to do with material procurement (e.g., buyers, material planners) should also be included in material overhead. Breaking out material overhead as a separate cost category is not a typical practice in the United States, like it is in Germany; however, doing so is recommended since there could be large swings in this cost between different types of materials. For example, a large part that is used only at low quantities might have to sit in inventory for a long time and then require a Hilo to transport, while a small high-volume part might only sit in inventory for a couple of days and need transport inside the plant using only carts. Other products might also require higher effort in sourcing and delivery assurance than others (e.g., sequencing and just-in-time [JIT]). Thus, on a percentage basis, some parts might have a material overhead cost of 10% of the purchased price, while others only 2%. These are significant swings on a percentage basis especially for companies that operate on low margins.

REAL-LIFE ANECDOTE

Tier 1 supplier of electronic assemblies was assuming the same material overhead cost percentage for every part regardless of type, size, or customer. However, during some investigation, it was discovered that 70% of its warehousing was dedicated to appliance customers who represented only 40% of its volume, which meant that this supplier was overcharging its other customers and undercharging appliance customers for material overhead. The supplier realized that this alone resulted in the appliance business being unprofitable and the other business very profitable, the opposite of what the supplier believed prior to the investigation.

MANUFACTURING LABOR

This is the cost for all the workers needed in order to produce the final product. These workers are usually split up into two main groupings, direct and indirect. Direct laborers are generally those involved in manufacturing processes that directly add value to the product. In most cases, these individuals operate the equipment or add value to the product manually (e.g., assembly). Often, individuals who touch the product during production but do not add value are also treated as "direct." This would include activities such as sorting or inspection. These are the operators that move the parts from one station to another or add components to the assembly or sometimes even operate the machines such as machining CNCs.

Indirect laborers are those who are not directly involved in manufacturing activities, but are there to support those activities. These include in-process material handlers, shift supervisors, machine technicians, maintenance technicians, manufacturing engineers, quality engineers, and all the other people indirectly engaged in the manufacture of a product. Some companies even include plant managers or accounting controllers and their staff in the indirect labor category; however, these individuals are not responsible for manufacture of any specific part, so they should be instead included in the manufacturing overhead cost category.

MANUFACTURING OVERHEAD

This is the cost at the manufacturing factory that is not directly tied to any specific product but rather to the operation of the factory's various processes. This would include the costs like building rent (if building is purchased by the company, then its cost would fall under general depreciation), maintenance, electricity, water, gas for the building and equipment, manufacturing supplies, any outside services such as uniform washing or cafeteria, and most likely the management and its staff (plant manager, controller, accountants, HR, IT, etc.).

It is important to point out that there are two types of manufacturing overhead cost, fixed and variable. The general rule of thumb is that fixed overhead is anything that would not change with production volumes, at least in the short- to mid-term. Things like building lease and plant management would be considered part of the fixed overhead cost. On the other hand, the variable overhead cost could go up and down with production volumes, meaning that as capacity utilization goes up and down, this cost would also go up and down. For example, equipment electricity is considered a variable overhead cost because it will not be used if there is no production.

SETUP

This cost category is often overlooked because the cost of setting up a tool or equipment or both is usually very small for large-volume products. However, this cost can be very large on a piece cost basis for small-volume products. This is because it might take couple of technicians a few hours to set up to run a part, which would cost the time of those technicians and the cost of opportunity lost by machines sitting idle (which could be worth thousands of dollars), while the batch run on a small-volume program might be only hundreds or a few thousand parts. The result of dividing thousands of dollars over only a few hundred or thousand parts is a potentially high cost per unit.

For example, if two technicians take five hours to set up a machine and they are paid $20 per hour while the machine cost per hour is $200, then the total cost of setting up is $1,200 (2 × 5 hours × $20/hour + 5 hours × $200/hour). If the batch of parts that will be run each time is only 2,400, then the cost per unit will be $0.50. If the rest of the part cost is only $0.25, then $0.50 in setup cost would be a huge impact on overall cost. Without considering this possible swing in cost, grave inaccuracies can be made in product costing.

EQUIPMENT DEPRECIATION

This is probably the most misunderstood cost category and not always considered separately from the manufacturing overhead category. In layman's terms, depreciation is a financial accounting concept that was designed to "smooth out" the company's annual investment amount over longer time periods, thus avoiding spikes/dips in the income in years with high or low investment. For example, if a company with annual sales of $100M and an annual profit of $5M wants to make a $7M investment in a particular year, instead of recording a $7M expense that year and thus wiping out all of its profit, it is given an ability to record that expense over a longer period of time (e.g., 10, 8, 7 years depending on type of equipment). This means that only a portion of that $7M investment is recorded for any given year (e.g., $700k if 10-year depreciation schedule is used), and the company can still record a profit for the year. However, another depreciation charge would be applied in the next number of years even if no investment is made during those years.

This technique was designed to protect the investors from incorrectly measuring a company's financial health. That's why today companies diligently keep track of all their depreciation with various assets at various depreciation time schedules. However, whether this method is helpful or not to accountants, it is really not an accurate measurement of cost. That is because any investment made in the past is already "sunk," meaning the money was already spent and thus it is irrelevant to any future decisions. What is important is the money that will need to be spent in the future on capital assets needed to maintain the company's existing capabilities. A company needs to accumulate the funds needed to replace an asset's capabilities as it uses that asset to produce and sell its products.

If any new investment has to be made for a new project or a company is still making payments on a previous investment, then these are the actual costs that will be incurred and should be accounted for in the product cost calculation. By including depreciation for equipment that a company already paid off, it is in effect inflating its cost and putting itself in position to fail in winning new business. On the other hand, a company with fully depreciated assets might fail to account for cost of maintaining production capability if it simply assumes no equipment depreciation cost or any other equipment cost. This will lead to winning business and then not having enough cash to cover various equipment costs.

SCRAP

There are many reasons why parts or materials are scrapped during manufacturing. Some material will be scrapped during the incoming inspection where it is deemed defective, some will be the result of breakdowns during the manufacturing process, and some of the parts will be scrapped after all the manufacturing processes are complete because they fail to meet quality standards.

Depending on a type of product and manufacturing process, the cost of scrapped parts can be significant. In the most extreme cases, like decorative painting for interior automotive components where there could be multiple paint applications and laser etching, the scrap can reach 30%. In other cases, scrap can be very low. A fully automated assembly process, for example, where all the steps are highly controlled and human error is eliminated, the scrap cost can be as low as 0.5%.

It is worth mentioning that scrap percentage sometimes can be misunderstood. Some manufacturing plants define scrap percentage differently than others. It is important to understand whether this percentage is of total cost or of total revenue or whether it reflects expected scrap during full production or scrap during production startup where it could be much higher.

Another important factor to consider is that scrap is cumulative (as shown in Table 2.1), meaning that scrap accumulates as parts go through the manufacturing sequence, and the amount of material used at each process step will be different at each stage. For example, a decorative plastic part will scrap only moldings in the molding process, but will scrap both moldings and paint during the painting process. So, although total scrap is important, a more accurate way to look at scrap is by understanding accumulation of scrap at each stage of the manufacturing process.

TABLE 2.1
Cumulative Scrap Calculation for Decorative Plastic Part

Process Name	Molding	Paint Layer #1	Paint Layer #2	Paint Layer #3	Pack
Batch volumes at each step	1,000	980	931	838	754
Resin cost per unit	$1.00	$1.00	$1.00	$1.00	$0.00
Paint cost per unit	$0.00	$0.30	$0.30	$0.30	$0.00
Total cost of processed resin	$1,000.00	$980.00	$931.00	$837.90	$0.00
Total cost of processed paint	$0.00	$294.00	$558.60	$754.11	$0.00
Total cost of material processed at each step	$1,000.00	$1,274.00	$1,489.60	$1,592.01	$0.00
Total cost of new material used at each step	$1,000.00	$294.00	$279.30	$251.37	$0.00
Scrap %	2.00%	5.00%	10.00%	10.00%	0.00%
Cost of scrapped resin	$20.00	$49.00	$93.10	$83.79	$0.00
Cost of scrapped paint	$0.00	$14.70	$55.86	$75.41	$0.00
Total scrap cost	$20.00	$63.70	$148.96	$159.20	$0.00
% cumulative scrap of total material cost	2%	6%	15%	21%	21%
% units scrapped			25%		
% total scrap cost			21%		

CORPORATE OVERHEAD

This is all the cost that is not directly related to manufacturing. For large companies, this would usually include all the activities at its corporate headquarters and not in its factories. This cost is often called the SG&A or selling, general, and administrative cost per the GAAP accounting standards in the United States (and similarly defined under the international IFRS accounting standards). What is typically included in this category is the cost covering the sales force and its marketing activities; any finance, accounting, HR, and IT staffs that are not related to manufacturing; the top management of the company; and the buildings and office equipment assigned to them.

For smaller companies, where the company headquarters are often in the same building as the plant, the separation between manufacturing and non-manufacturing related staff is often more difficult because everyone's roles seem to blend together. However, those companies would be encouraged to at least roughly make that separation using activity-based costing, which would help properly allocate cost.

It is worth noting that there will be slight differences from country to country in how some costs are allocated. For example, in some countries (e.g., Germany), the corporate overhead may also include the management staff at the manufacturing factory. Another difference might be in where R&D (research and development) is included. Under GAAP reporting rules, R&D is often included in the SG&A cost. However, under IFRS, R&D is often reported as a separate cost bucket.

A company might also use two different methods to keep track of cost, one for purely accounting purposes and one for costing or quoting purposes. It is important to understand the distinctions between the two methods when developing cost estimates for quotes to customers or when analyzing suppliers' costs.

PROFIT

This is simply the money earned by a company by making a product and is the difference between its sales price and all of its costs. As straightforward as this sounds, there are many confusing terms that try to quantify a company's profit. Following are some of the most common:

Gross Margin or Gross Profit – company's total sales revenue minus its cost of goods sold (COGS) (not including corporate overhead cost), divided by total sales revenue, expressed as a percentage. The gross margin represents the percent of total sales revenue that the company retains after incurring the direct costs associated with producing the goods and services it sells. The higher the percentage, the more the company retains on each dollar of sales to service its other costs and debt obligations.

Operating Profit or EBIT (Earnings Before Interest and Taxes) – profitability of the business, before taking into account interest and taxes. To determine operating profit, operating expenses (corporate overhead) are subtracted from gross profit. Operating profit is a key number for managers to watch as it reflects the revenue and expenses that they can control.

TABLE 2.2

Sample Financial Calculation

Revenue	$100
Cost of goods sold	$50
Gross profit	$50
Operating expenses	$30
Operating profit, EBIT	$20
Interest	$3
Taxes	$5
Net profit	$12

EBITDA (Earnings Before Interest, Taxes, Depreciation, and Amortization) – similar to EBIT but it also excludes depreciation and amortization as a proxy for cash flow indicator.

Net Profit – also referred to as the bottom line, net income, or net earnings is a measure of the profitability of a venture after accounting for all costs and taxes. It is the actual profit, and includes the operating expenses that are excluded from gross profit.

Table 2.2 shows a sample calculation indicating the difference between some of the financial terms.

PACKAGING (OUTBOUND)

Same considerations as inbound packaging, but this is packaging cost for a product that is leaving a supplier bound for customer.

LOGISTICS (OUTBOUND)

Same considerations as inbound logistics, but this is logistics cost for a product that is leaving a supplier bound for customer.

TOOLING AND FIXTURES

In order to produce any part, a tool or a fixture is needed to perform almost every manufacturing operation. A casting or a molding will require two separate steel plates, each with an opposite half of the part shape machined into the plates. A machined part often will require multiple diamond tipped tools that remove material from parts. Other processes, like painting or custom assembly, will require fixtures to hold parts in place during the manufacturing process.

The cost of tools and fixtures can often be very significant, so it is important as a company to have the capability to estimate this cost. There are three basic tool cost categories that comprise tooling cost: design, raw materials and purchased components, and tool/fixture manufacture. Typically, design and manufacture are defined in cost per hour, which can vary widely depending on type of tool/fixture, country of design/manufacture, and part complexity.

It is also important to note that a tool life can vary dramatically depending on type of manufacturing process and tool materials chosen. A typical injection molding tool could have a life of one million cycles (moldings) while a die casting tool life could be only one hundred thousand cycles. It is critical for the customer to understand if the purchased tools have a long enough life to last the full program life and if not, then how many additional tools are required for either purchase or amortization into piece price.

CONNECTOR EXAMPLE

To provide a clearer description of the various cost categories discussed earlier, described below is a cost estimating example for a simple connector (Figure 2.2) that can be used in the SPARK hybrid truck to connect electronic devices.

This particular connector has six pins, each with an attached wire. Let's assume the pins are over-molded by a plastic housing with its value chain shown in Figure 2.3.

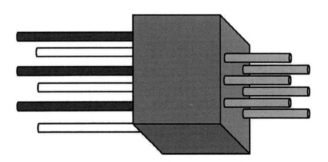

FIGURE 2.2 Sample six-pin connector.

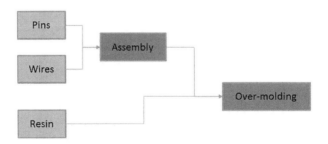

FIGURE 2.3 Connector value chain.

Based on the value chain, a cost estimator at the supplier making this connector will estimate its cost and prepare a piece price quote to SPARK. At the same time, a cost estimator at SPARK will try to estimate what the connector should cost. Both of them will, or should, follow a cost estimating methodology as described below.

RAW MATERIAL

For the connector manufacturer, resin will be the only raw material required to mold its housing. The pins and wires will be purchased in final form. After reviewing the annual volume requirements (10,000 vehicles per year with one connector per vehicle means a part volume of also 10,000), the supplier's cost estimator assumes that this part will be manufactured in a two-cavity molding tool, meaning that two parts will be molded at the same time using the same tool, and a 60-ton molding press will be needed.

In addition to the housing net weight (weight of the finished housing), the estimator will also have to account for all the resin that will be discarded, such as runner, sprue, and offal (flash). The weight of the part, including all the discarded resin, is called a gross weight. Figure 2.4 shows an example of a two-cavity mold and its various components.

The net weight is usually specified in the engineering drawing, which in the case of the connector is 10 grams. The gross weight, on the other hand, can only be accurately defined after the mold flow analysis is completed by the design and/or manufacturing engineer. In our connector's case, it is too early in the product development process for the mold flow analysis to be completed, so the estimator will have to use his manufacturing experience to estimate the gross weight.

For this particular part and cavitation (number of cavities per mold), the supplier's estimator assumes an additional 5 grams of resin that will be discarded, thus bringing the total gross weight of a part to 15 grams (10 grams net weight + 5 grams discarded weight). For the two parts being molded in one tool, the total "shot" weight

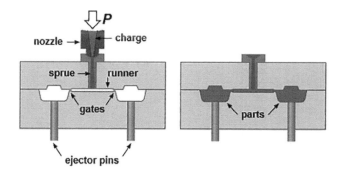

FIGURE 2.4 Two-cavity mold. (Reproduced courtesy of ariel cornejo and Wikimedia Commons; https://commons.wikimedia.org/wiki/File:Injection_molding_diagram.svg.)

will be 30 grams. This means that the total amount of resin used, or shot into the tool, to make two parts is 30 grams.

Keep in mind that the part design often allows for the discarded resin to be ground down to pellet size and reused; therefore, the cost of that resin can be discounted. Even in cases where the discarded resin cannot be reused, a molding supplier can often sell the reground resin at a discounted price. For simplicity's sake, let's assume that neither is done for our connector.

The engineering drawing calls out a resin that is 30% glass filled, which typically can be purchased in the market for $2.50 per kilogram. This assumes a certain amount of it being purchased per year since the price is heavily dependent on amounts purchased. Thus, the cost of raw material is calculated by multiplying the part gross weight of 15 grams by the cost per gram of $2.50/1000 grams, or $0.0025 per gram, which results in a cost of $0.0375 per part.

The customer's cost estimator follows the same method of calculation, but since the supplier's internal cost information is not available, he/she assumes that only 26 grams of resin per shot will be used and that the best possible resin price of $2.00/kg will be achieved. These assumptions lead to a per piece cost of $0.0260 (13 grams × $2.00/1000 grams).

Although these calculations by both estimators are very detailed, it is worth noting that there were several assumptions made along the way that may or may not be questioned. The first question could be, "How did the cost estimators know what amount of additional resin would be needed for runners, sprues, gates, offal, etc.?" As was mentioned before, a cost estimator would spend years gaining experience in both cost estimating techniques and the molding manufacturing process. Ideally, he/she would already have had worked on a similar part, so that even without manufacturing process or tools designed yet, he/she would be able to make those assumptions with reasonable accuracy. The second question might be, "How did the cost estimator know the price of that particular resin to be $2.50 or $2.00 per kilogram?" This also comes with experience and research. There is a lot of information available publicly on resin pricing from indexes or resin suppliers themselves. A cost estimator would also make an adjustment to the resin price based on the amount of resin purchased annually by a specific connector supplier. If the resin is unique for this particular supplier, the annual volume purchased would most likely be low, resulting in a higher price of resin. If another connector supplier is already molding similar parts and is buying this particular resin at very high annual quantities, the resin price would probably be lower.

Purchased Components

In this case, the connector supplier will be purchasing pre-made pins (stamped and plated by another supplier) and wires (stranded copper wires in a plastic jacket assembled by another supplier). For both pins and wires, a cost estimator might choose to follow the same type of costing analysis as with in-house manufactured components: estimating each element from the ground up. Since supplier and customer cost estimators might not have much experience estimating these parts, they will need to use other data for their cost assumptions.

The supplier's cost estimator asks the purchasing department to send the part designs out for quotes to their sub-suppliers, which come back at $0.03 and $0.04 per pin and wire, respectively. On the other hand, the customer's cost estimator relies on historical data from previous quotes, which are $0.03 per pin and 0.05 per wire. So, for our six-pin connector, our total purchased costs are $0.42 (6 × $0.03 + 6 × $0.04) per supplier's cost estimate and $0.48 (6 × $0.03 + 6 × $0.05) per customer's cost estimate.

It is important to note that historical cost data can be a good source of purchased cost information; however, it should be validated at some point using cost estimating methodology. It is also possible to develop cost models that can predict component pricing based on some basic inputs, such as weigh, length, volume, etc.

PACKAGING (INBOUND)

The packaging for resin is typically already included in the resin price, so the only packaging cost for our connector would be for the purchased components, pins and wires. Since a cost estimator will not have detailed packaging requirements this early in the product development process, he/she will have to assume something based on historical data. Let's assume that based on several data points, packaging cost is usually 1% of total purchased component cost, so $0.0042 ($0.42 × 1%) and $0.0048 ($0.48 × 1%) for the supplier's and customer's estimates of our connector, respectively. Since packaging is a very small percentage of overall cost and inaccuracy in cost assumption has minimal impact on profitability, this assumption is accurate enough.

LOGISTICS (INBOUND)

A supplier that will be making this connector will have to bring in raw material and purchased components to its manufacturing facility. Their estimator might already know where these will be purchased and will have some historical cost data from the company's logistics department, which in this case is $0.0210 per part.

The cost estimator at the SPARK hybrid truck company will most likely not have this information, so an assumption will have to be made about rough location of suppliers. Let's assume that based on some historical data a typical logistics cost is anywhere between 2% and 5% of material cost, depending on the level of supply base localization. Since the overall material cost is $0.5060 ($0.0260 + $0.48), then the minimum cost of logistics at 2% (which assumes all suppliers are localized) is $0.0101.

It is important to note that resin logistics cost is often included in the price of resin. If that is the case, then there would be no need to add additional 2% on top of the $0.0260 resin cost.

MATERIAL OVERHEAD

Most companies do not identify material overhead as one of their cost centers or cost "buckets," but it is usually a significant enough cost to consider separately for accuracy's sake. Simply spreading these costs across other cost buckets like "peanut butter on toast" will cause an inaccurate price calculation that might lead to a bad company decision. This is because some parts will have much higher material overhead costs than others, and an incorrect assumption will skew the cost estimate in the wrong direction by possibly a large amount. Since the warehouse inside a manufacturing facility could take up to a third or more of the overall space and employ several buyers, material planners, and material handlers, as well as various moving equipment and storage racks, this would be a significant cost to the company.

The supplier's cost estimator will have a benefit of having the internal historical data or at least access to the data needed to establish these costs; however, the cost estimator at the SPARK hybrid truck company will have no such access, thus a good starting point for material overhead might be 5% of the total material cost. If the cost estimator feels that a particular part will require a disproportionate amount of storage space or time to procure and handle it, then he/she can increase that amount up to perhaps 10% or more.

A specific case where the material overhead cost will be unusually high is when the part volumes are low but a company orders large quantities in order to get a reasonable price. This would mean that the parts will sit on the shelf for longer than normal. Another example is when material planners and/or buyers must spend more time procuring and ordering parts due to sub-supplier issues (e.g., lack of supply).

For our connector, let's assume that low volumes and difficulties in procuring material will result in a high material overhead cost per part. The cost estimator at the connector supplier will have a good chance to understand the cost impact based on actual cost seen at their manufacturing plant, which in this case is $1,500 per year or $0.15 per part ($1,500/10,000 parts).

The cost estimator at SPARK will have to make an estimate based purely on some previous experience. Let's assume that he/she assumes 10% material overhead for the connector or $0.0506 per part ($0.5060 × 10%). As a "sanity check," the customer's cost estimator might want to roughly evaluate if the total annual dollar amount spent on material-related activities makes sense. For example, if the annual connector volume is 10,000, then the annual material overhead cost for this connector will be $506 (10,000 × $0.0506). If a person employed for this activity makes $10 per hour, then this means that about 50 hours annually ($506/10) are spent to manage the material activities on this connector. Is that enough? If this connector is used for the SPARK application and there is no other volume to spread the costs over, this may not be enough money to cover all the real cost associated with managing the material. In this case, the customer's cost estimator assumes that $506 per year will be enough.

Manufacturing Labor

In a later chapter on cost allocation, we will discuss activity-based costing and the importance of assigning as much cost as possible directly to the product. For the manufacturing labor especially, this means allocating both direct and indirect employees to the manufacturing process needed for a given product. Although many companies often bundle indirect labor into manufacturing overhead, only to spread it across all products by the same percentage indiscriminately, this is often inaccurate and misguided. Both should be allocated on their own merit, so direct and indirect labor costs will be explained separately.

DIRECT LABOR

As was defined earlier in this chapter, the direct labor cost accounts for all the workers that provide manual labor to the product directly during its manufacturing process. In the case of the connector, there are two main manufacturing processes, assembly of wires to pins and over-molding of that assembly with a housing.

For the assembly process, the supplier's cost estimator determined that this will be a manual operation that will take a laborer roughly 15 seconds to complete for each assembly on a simple bench with a holding fixture. The parts will be assembled at the company's southern Mexico plant where the labor rate is $3 per hour. However, the company must spend on average an additional $1.50 per hour to cover the cost of fringe benefits that each laborer receives, such as health insurance, bus pickup, free lunches, training, etc. That means that a fully fringed (including benefits) labor rate is $4.50. Also, the company pays its laborers for lunch time and breaks (about one hour in an eight-hour shift) plus for any unplanned machine downtime. Both unplanned and planned downtime, often called inefficiency, take up about 15% of laborers' eight-hour shift (or 72 minutes). Note that a reverse of inefficiency is often called the efficiency factor, which in this case would be 85%. So, the direct labor cost for the assembly operation calculates to be $0.0215 (15 sec/3600 sec × 1 operator × $4.50/hr × 1.15 inefficiency).

Similarly, for the over-molding process, the supplier's cost estimator determined that one laborer will be needed to operate a 60-ton molding press. The operator will place the assembled pins with wires into the molding tool, then take the over-molded finished part out of the molding tool. The whole process will take about 30 seconds per mold, which has two cavities, meaning that two parts will be over-molded each time. This means that a cycle time per part, or the time to manufacture one part, is roughly 15 seconds (30 seconds/2 cavities). Inefficiency for this process can also be assumed to be around 15%, thus the direct labor cost for the over-molding process also calculates to $0.0215 (15sec/3600sec × 1 operator × $4.50/hr × 1.15 inefficiency).

In the case of the customer's cost estimator, he/she will not have the benefit of having dedicated manufacturing engineers to define the supplier's manufacturing process, so the assumptions will have to be more of a guess if there is no historical data available. The customer's cost estimator assumes that each process will take only 13 seconds per part with only 1.5 operators total and an efficiency of 90%. This results in total direct labor cost of $0.0268 (13 sec/3600 sec × 1.5 operator × $4.50/hr × 1.10 inefficiency).

It is important to note that reducing cycle time does not always mean a reduction in cost. For example, if the cycle time to make a connector is reduced from 15 seconds currently to 13 seconds, our calculated direct labor cost for each operation would decrease to $0.0187 (13 sec/3600 sec × 1 operator × $4.50/hr × 1.15 inefficiency). However, for those two seconds saved for each part or 480 seconds per hour (at full utilization), would the operator be dismissed or would a shift of operators be eliminated or would another product be produced? In this case, the answer to all these questions is no, because a reduction of two seconds per part only results in one hour saved per shift. For that one hour per shift or three hours per day (assuming

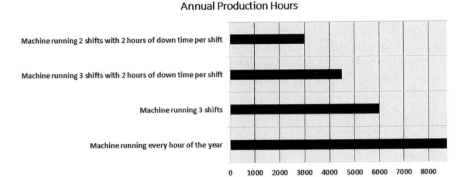

FIGURE 2.5   Annual equipment availability.

three eight-hour shifts), no workers can be eliminated. Each shift will remain running at full employment for the full 24 hours per day. Thus, the cost to make each part will not change.

Since inefficiency or efficiency factors are often confused with OEE (overall equipment effectiveness) or capacity utilization or uptime, it is important to make the differences clear. Inefficiency is the time lost due to planned and unplanned downtime during manufacturing. Uptime is the opposite of that, just like efficiency factor, and represents the time a machine was actually running. OEE is similar but it also discounts the time lost on making bad quality or scrapped parts. Capacity utilization also accounts for time lost due to lack of business, meaning that if only two eight-hour shifts are running, the eight hours on the third shift are lost and thus the capacity utilization is only two thirds or 66.67%. Figure 2.5 shows the importance of optimizing machine utilization. The ideal state is to have the machine run 24/7 (24 hours a day and 7 days a week) every week of the year because the more machines run, the more parts that they can make, the more money can be made covering the factory cost, and ultimately the more profit can be generated. Machine utilization and corresponding profit potential are reduced dramatically with fewer shifts and various downtime issues.

INDIRECT LABOR

It is often helpful to visualize a manufacturing plant when allocating cost to activities and people that do not directly touch the product. It is easier to do so for a cost estimator at the supplier who can simply walk out on the manufacturing floor and observe. The customer's cost estimator does not have that luxury; therefore, it is important that he/she visits as many supplier manufacturing plants as possible to obtain an understanding of typical and best in class manufacturing processes.

For a connector supplier, a typical manufacturing plant layout might be something like the one in Figure 2.6. This plant can be split up into four distinct

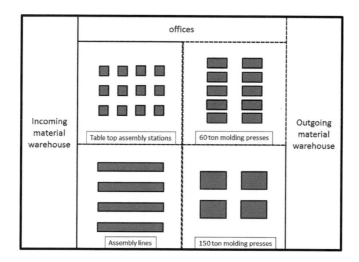

FIGURE 2.6   Connector manufacturing plant layout.

departments: table top assembly stations, assembly lines, 60-ton press molding, and 150-ton press molding. Under this setup, it might be typical to have a shift leader for each department and manufacturing and quality engineers for each assembly and molding areas. There would also be a couple of technicians shared between all four departments.

In the case of our connector, which would be produced in a single table top assembly station and over-molded on a small 60-ton press, there would be cost for two shift leaders ($6/hour each), one half of manufacturing engineer and one half of quality engineer ($16/hour each), plus one technician ($10/hour each). So, the total cost of indirect labor per hour is $48 (2 × $6 + 2 × $16/2 + 1 × $10). However, this $48 also covers production of other parts, not just our connectors. If no real data are available, the total production can be estimated. Since there are 12 single assembly stations and 12 small 60-ton presses, if we roughly assume that all parts are made in 15 seconds each, then 2,880 parts are made per hour (240/hour × 12). This means that the $48 in indirect hourly costs will be spread over 2,880 parts, which is $0.0167 per part.

On the other hand, the customer's cost estimator will usually have less information since he/she is not always able to view the supplier manufacturing plant and processes. In this case, a cost estimator might assume that indirect labor cost is some percentage of the direct labor cost, perhaps 50%, which in the case of our connector results in $0.0134 per part. This is just an assumption and a starting point for a later negotiation.

SETUP

As mentioned before, setting up for a production run is a fairly significant cost per part in cases when the production runs are very short. This is typical of low-volume products since, with the advent of lean manufacturing, companies prefer making only the amounts needed for immediate production in order to avoid holding inventory (cash on a shelf). For example, the connector supplier will produce parts only once per week for a required weekly shipment to the customer SPARK. That means only 200 parts will be produced each week (10,000/year/50 weeks), which will require only less than one hour of production per week. However, the time to set up for the assembly process is half an hour by one technician (with $6/hour labor rate), and the time to set up the molding press is one hour by two technicians (with $8/hour labor rate). Thus, the total setup cost is $3 and $16 for assembly and molding, respectively. This is just the labor cost, but there is also lost machine time or time in which the machine was stopped for setup while it could have been running production. For simplicity, we will not include lost machine time cost here. The $19 in setup cost has to now be absorbed by 200 parts that are produced during each run, so the total setup cost per part is $0.095 ($19/200 parts).

On the other hand, the customer's cost estimator assumes that the supplier will make the parts in only one run per year, store the parts in a warehouse, and then ship weekly to the customer. The cost estimator also completely ignores the added cost of inventory holding assuming that manufacturing or material overhead covers this extra cost. Thus, per customer's cost estimator, the setup cost is only $0.0019 per part ($19/10,000 parts).

Manufacturing Overhead

This cost is not difficult to measure, but it is extremely difficult to allocate to specific products. This makes it by far the most hotly contested cost category between customers and suppliers. For example, how much time does a plant manager, who is part of manufacturing overhead, spend on any one product? This is difficult to quantify and might change continuously depending on any given day in a manufacturing plant. How do you allocate office space at a plant to any specific product? Office space is used to house accountants, IT administrators, HR professionals, and such, but their jobs and offices that they use are not product specific.

Instead of trying to allocate to products, what companies often do is find shortcuts. A company might, for example, look at the previous year's total factory costs and find that for every dollar that was spent on labor, another three were spent for expenses related to manufacturing overhead. Therefore, this company might decide to estimate all future manufacturing overhead costs using the same three-to-one ratio. If it is quoting a product that requires one dollar of labor, it will assume that three dollars of manufacturing overhead cost will be needed. Its simplicity makes it a very common method, but, as we will discuss later, it can be very inaccurate.

Other companies try to develop machine rates for every piece of manufacturing equipment in the plant. However, this method of allocation is often based on the original equipment cost of the machines, which means that, similar to the labor to overhead cost ratio, each dollar of capital investment gets the same ratio of overhead expenses. For example, in the case of the connector supplier, since the molding presses are a lot more expensive than table top assembly stations (~$200,000 versus ~$10,000), the presses would absorb the bulk of all the manufacturing overhead cost. Again, this may or may not be an accurate assumption, as we will discuss later.

Yet another method used by companies is to simply check their historical manufacturing overhead as percentage of revenue, then apply the same markup percentage to every product that they make. For example, if for the company as a whole that percentage happens to be 20, then the company would multiply the product cost (e.g., material and labor) by 1.20 to get to the price. Unfortunately, this is usually the most inaccurate way of estimating manufacturing overhead cost, because it completely ignores manufacturing complexities that might exist in the company's product portfolio or volume mix of those products.

The most accurate method of cost allocation is activity-based costing (ABC), which will be discussed in more detail later in this book. As its name suggests, this method identifies specific activities associated with manufacturing overhead, then uses drivers for each activity to allocate costs to specific products. For example, perhaps it could be assumed that since the plant manager and most of his/her staff are there to manage people and not products, then a driver to allocate that cost could be the number of man hours spent on each product. However, other costs, such as those associated with electricity and water, could be allocated to specific machines and converted into hourly rates.

In the case of the connector, let's assume that the supplier cost estimator used ABC methodology and developed machine rates for the two manufacturing processes at \$15/hour and \$20/hour for assembly and molding, respectively. This results in the manufacturing overhead cost of \$0.1677 per part (15 sec/3600 sec × (\$15 + \$20)/hr × 1.15 inefficiency).

The customer cost estimator does not have access to the supplier's data and allocated cost based on original equipment cost, which resulted in \$5 and \$30 per hour for assembly and molding, respectively. This results in the unit cost of \$0.1453 (13 sec/3600 sec × (\$5 + \$30)/hr × 1.15 inefficiency).

It is important to point out that since the cycle time assumptions were the same for both processes and the connector required both assembly and molding, the hourly cost of the two separately does not really matter in our example. This is because both added up to \$35 total cost per hour (both processes could be thought of as a system that a part has to go through). On the other hand, if the connector required only assembly or only molding, then the method of allocating manufacturing overhead by original cost of equipment would have been inaccurate. The reason for this will be discussed in more detail later in the book, but it is because the cost of running an assembly operation for the connector plant is almost as high as for the molding. Using the cost of original equipment as an allocation driver would skew the cost incorrectly.

DEPRECIATION

As was mentioned before, the concept of depreciation is purely an accounting invention and it perhaps serves its purpose there. Nevertheless, most customer estimators would typically calculate depreciation using the original cost of the equipment, perhaps \$10,000 and \$200,000 for assembly and molding equipment, then amortize it over 8 and 10 years, respectively, and then over total annual production hours of that equipment. This would result in hourly rates of

$0.25/hr ($10,000/8 years/5,000/year) and $4.00/hr ($200,000/10 years/5,000/year) for assembly and molding, respectively. The depreciation cost for each unit produced would be $0.0019 (15/3600 × $0.25/hr × 1.15) for assembly and $0.0192 (15/3600 × $4/hr × 1.15) for molding.

To understand real cost of manufacturing, the supplier's cost estimator might want to research his/her company's capital investment costs in more detail. For example, what if the assembly stations were bought with cash from a bankrupt company for $1,000 per station and the company doesn't foresee replacing any of the stations in the next 10 years? This means that the cost incurred in the past for the 12 stations is only $12,000 and that no additional cost will be incurred in the future. As such, if a company required a payback for actual investment, the cost estimator would need to account for $12,000, not for $120,000 as the original equipment value of $10,000 per station would suggest. In reality, if the company chose to use a typical depreciation calculation for their customers, the additional $108,000 ($120,000–$12,000) of depreciation is really just additional profit that the company would receive during the life of the program (assuming they won the bid). It is not a real cost. The actual hourly cost for each assembly station is $0.02 ($12,000/10 years/5,000 hours/12 stations).

Similarly, the supplier's cost estimator might find that the 60-ton molding presses were purchased used on a loan and that the company expects to replace two of the older machines in the next 10 years. Therefore, assuming that the outstanding loan amount (principal and interest) is $1,000,000 and the two replacement machines will cost $200,000, the company will have a capital investment cost of $1,200,000 over the next 10 years for the 60-ton molding press department. This results in an hourly machine investment cost of $2.00 per machine ($1,200,000/10 years/5,000 hours/12 machines).

This means that our connector will have a depreciation, or capital investment, cost of $0.0111 (15/3600 × ($0.023 + $2.30) × 1.15). This is 50% less than it would be if a pure accounting depreciation was used.

The supplier's cost estimator also confers with the manufacturing engineer and is told that for this specific project, $10,000 will have to be invested in order to make repairs to one of the molding presses. This does not change the machine's hourly depreciation rate, but the new investment will have to be considered as part of the ROI calculation and thus will change the investment cost per part for this particular project. This will be explained later in this book.

The cost of equipment maintenance and repair should also be part of the total equipment cost. For example, the molding press department may need a dedicated maintenance technician and about $30,000 in replacement parts per year, which would result in $1.16 hourly cost (($40,000 salary + $30,000 for parts)/12 machines/5,000 hours) or $0.0055 per part (15/3600 × $1.16 × 1.15). Increases to the annual repair cost of equipment should also be considered with the equipment's increasing age.

The above suggested calculation was possible for the supplier's cost estimator, because he/she has the actual data. For the customer's cost estimator, who does not have such luxury, the best initial approach might be to follow the typical depreciation calculation method. This should be a good enough starting point in the cost engineering process and can be refined later in the negotiation process.

SCRAP

This cost is inherent to the manufacturing processes and products being made. In the case of our connector, both supplier and customer cost estimators feel safe to assume, based on historical data, that scrap cost is equivalent to 2% of overall incurred cost. This means that not only material cost, but also processing cost, will have to be scrapped as both are wasted on production of bad parts. Some companies choose to account for the manufacturing process waste in the inefficiency factor and some do it in manufacturing overhead. It really does not matter where this cost is bucketed; the key is to make sure it is accounted for somewhere and distributed to products accurately.

In the case of our connector, both estimators assumed 2% of cost is scrapped, but their total cost is different. This results in $0.0194 and $0.0160 for supplier and customer cost estimators, respectively.

It is important to note that for some processes, such as painting, it will be necessary to develop a cumulative scrap calculation where material scrap is calculated after each process. For example, let's consider the case of an automotive Class A bezel for the radio and HVAC control center. This part is first injection molded, then pad printed, then painted, then laser etched, and material scrapped at each process might be 2%, 5%, 10%, and 3%, respectively. The material scrapped at each stage will have different levels of both material and manufacturing cost content. Those will have to be added to fully understand the total cost of scrapped material.

CORPORATE OVERHEAD (SG&A)

Just like manufacturing overhead, corporate overhead cost is difficult to allocate to specific products, because corporate activities cover all products and sometimes they have nothing to do with products at all. For example, the salary of the company's CEO would be difficult to allocate to any specific product if a company makes a hundred or a thousand different products. This is why, usually, cost estimators assign the same margin percentage for all products on top of manufacturing cost. This percentage is usually based on historical data from previous year's profit & loss (P&L) statement. For example, if last year's revenue was $10 million and the total spend on corporate overhead (SG&A) was $1 million, then cost estimators would assume 10% as their corporate overhead.

However, allocating corporate overhead cost for every product based on the same percentage can be very inaccurate and dangerous to company's well-being. If the same percentage is applied to high- and low-volume projects, then most likely the low-volume projects will be under costed since on an absolute basis, there will not be enough cost allocated to cover all the required corporate expenses. On the other hand, the high-volume products might be over costed because they are incorrectly absorbing cost from low-volume products.

For example, let's assume that a company assigns the same 10% corporate overhead on two quoted connector projects, one worth $50 million and the other $5,000 per year in sales. The 10% assumed for both means that the first project will have $5 million allocated in annual corporate overhead cost, while the second will have only $500 allocated. Considering that the company's total SG&A cost last year was $1 million and that based on some internal calculations it was found that each project, regardless of volume, needs minimum of $5,000 annually to support; it is very likely that allocating $5 million is too much for the first project and $500 is too little for the second project.

Although many believe that applying a flat percentage on all projects will eventually even out, meaning that some high-volume projects will pay for the under costed low-volume projects, what happens in reality is that a company that assumes a flat percentage will lose projects that it prices too high and will win a lot of projects that it prices too low. Eventually, this company will end up with mostly low-volume projects that cost much more than originally assumed and are most likely to lose money.

For our connector example, let's assume that the supplier's cost estimator is aware of the minimum annual corporate cost required to support a low-volume project and calculates the corporate overhead cost to be $0.50 per part ($5,000/10,000 volume). On the other hand, the customer's cost estimator uses 10% markup across all products for simplicity and arrives at $0.0818 per part (all mfg. cost $\times$ 10%). Note that the customer's estimator also incorrectly used markup $(X \times (1 + 0.10))$ instead of margin $(X/(1-0.10))$, which is a common mistake.

Profit

Much like the corporate overhead, most companies simply apply a profit margin percentage on top of all the cost. This is also usually based on historical profit levels, market competitiveness, or the company's profit requirements. However, just like with corporate overhead, assigning the same profit percentage to every project can be highly inaccurate and can lead to some bad decisions by the company. Using the same example as the corporate overhead, what if the same 10% profit margin was applied to two projects, one worth $50 million and one worth $5,000 in revenue. Again, this would result in $5 million and $500 in annual profit, respectively.

An owner of a company that, let's say, made $1.5 million in profit the previous year should look at these and be able to decide that an additional $5 million in profit would be nice, but is not necessary if it means losing a project. On the other hand, for the smaller project, $500 in profit is probably not worth working on, and an owner would probably pass on it.

Fortunately, there is a better way of assessing worthiness of a project than a flat profit percentage, and that is the ROI and the payback calculations. These methods ask the basic question, "If I invested X amount of money, how much money could I get in return and how fast?" For our two projects, what if the new investment was $500,000 and $10,000 for the large and small projects, respectively? This means that, for the larger project, the company could get its money back in just one year if it assumed an annual profit of $500,000 (or just 1% margin) and no financing required. This gives the company flexibility in pricing this project and can assume as little as 1% profit or even less if it chooses to get paid back over a longer period of time.

On the other hand, the smaller project would take 20 years to pay back the $10,000 investment if a flat 10% profit was assumed ($10,000/$500 annual profit). This means that the company, if at all interested in such a low-volume project, would need to increase the profit to perhaps $10,000 annually to get paid back in one year for their investment, if that is desired.

For our connector example, the supplier's cost estimator understands that his/her company's requirement is to get paid back for its investment within one year for any small-volume projects, therefore he/she calculates the profit to be $1 per part ($10,000 investment/10,000 volume).

In contrast, the customer's cost estimator assumes 10% profit markup for all projects and does not take ROI into consideration. Thus, his/her calculation results again in $0.0818 per part (all mfg. cost × 10%). Note that the estimator again incorrectly used markup $(X \times (1 + 0.10))$ instead of margin $(X/(1 - 0.10))$, which is a common mistake.

PACKAGING (OUTBOUND)

After the product is manufactured by a supplier, it is then shipped to the customer. In order to do so, the product has to be packaged, which usually consists of cardboard boxes, then put on a wooden pallet and probably shrink wrapped. However, the product could also be placed in plastic totes, which the customer would later return to the supplier (this kind of packaging is often called returnable, while cardboard packaging is often called expendable). The cost of either packaging solution is typically amortized in the unit price of each part.

In the connector example, the supplier's cost estimator knows that perishable cardboard boxes will be used with wooden pallet and shrink wrap, which will cost $100 per pallet. He/she also learns from the company's packaging expert that each pallet will be able to hold 2,000 parts, which means that the packaging cost per part is $0.05 ($100/2,000 parts).

On the other hand, the customer's cost estimator does not have the benefit of having a packaging expert to help and uses historical data that suggest an average packaging cost that is 2% of part price, which in this case results in $0.0196 per part (price × 2%).

LOGISTICS (OUTBOUND)

Often times, the customer chooses to pick up finished parts from the supplier, but just as often it is the supplier's responsibility to ship and pay for shipment. Determining cost of freight and customs duty can be very complex due to ever-changing shipping rates and tariffs applied by various countries. So, the supplier's cost estimator gets a shipping quote of $0.03 per part from their internal logistics expert.

The customer's cost estimator does not have easy access to logistics experts and uses historical data that suggest an average logistics cost that is 3% of part price, which results in $0.0295 per part (price × 3%).

SUMMARY OF CONNECTOR ESTIMATE EXAMPLE

To summarize the calculations for all the cost categories, Table 2.3 has costs and assumptions for both supplier and customer cost estimator.

TABLE 2.3
Comparison of Connector Estimates

	Supplier Cost Breakdown to Customer	Supplier Assumptions	Customer Cost Estimate	Customer Assumptions
Raw material	$0.0375	15 grams/part, $2.50/kg	$0.0260	13 grams/part, $2.00/kg
Purchased parts	$0.4200	Sub-supplier quotes	$0.4800	Historical data
Packaging	$0.0042	1% of material cost	$0.0048	1% of material cost
Inbound logistics	$0.0210	Logistics quote	$0.0101	2% of material cost
Material overhead	$0.1500	Historical data	$0.0506	10% of material cost
Direct labor	$0.0430	15 sec/part, 2 ops, 85% efficiency	$0.0268	13 sec/part, 1.5 ops, 90% efficiency
Indirect labor	$0.0167	$48 per hour	$0.0134	50% of direct labor cost
Setup	$0.0950	$19 per setup, 200 parts per run	$0.0019	$19 per setup, 10,000 parts per run
Manufacturing overhead	$0.1677	Assy: $15/hr, Mold: $20/hr	$0.1453	Assy: $5/hr, Mold: $30/hr
Depreciation	$0.0166	Assy: $0.023/hr, Mold: $2.30/hr	$0.0211	Assy: $0.25/hr, Mold: $4/hr
Scrap	$0.0194	2% of incurred cost	$0.0156	2% of incurred cost
Total Manufacturing Cost	*$0.9911*		*$0.7956*	
Corporate overhead (SG&A)	$0.5000	$5,000 cost per year	$0.0796	10% markup
Profit	$1.0000	1 year payback on $10k investment	$0.0796	10% markup
Price	*$2.4911*		*$0.9547*	
Outbound packaging	$0.0500	$100 per 1,000 parts per pallet	$0.0191	2% of price
Outbound logistics	$0.0300	Shipping supplier quote	$0.0286	3% of price
Price w/ Packaging and Logistics	*$2.5711*		*$1.0025*	

As is the case with all estimates, there are many differences between the supplier's and customer's cost estimates. In fact, there would be differences between estimates if two different estimators at supplier or customer estimated costs for the same part. This is why estimates are called estimates and not "exactomates." No estimate is ever exact; it only hopes to be the most accurate. Although some people would shudder at the thought, the truth is that most estimates can only achieve accuracy within 5% to 10% of actual. With transparency and collaboration, however, the estimate accuracy will improve during product development as design, manufacturing process, value chain, and other assumptions are fine tuned.

In the case of our connector, the differences or the gaps between supplier's and customer's estimates are very large (estimated price by supplier of $2.5711 versus $1.0025 customer's estimate). As the supplier and customer sit down for the first time to agree on a reasonable price, they will have to explore the gaps further. This can be a long and complex process with each side trying to get the best price for themselves often regardless of true cost, which may or may not be revealed to the other side. This cost-based or fact-based negotiation is much different than a typical collect and compare technique, where a buyer chooses the best quote out of many. A separate chapter will be devoted to cost-based negotiation. However, there is also a business case consideration that can make this negotiation even more complicated.

BUSINESS CASE CONSIDERATION

Most cost estimators prefer to simplify their method of calculations, thus "point in time" estimating methodology described earlier for the connector is very popular in the manufacturing industry. This means that a cost estimator looks only at today's or past cost of a product and does not consider any future cost fluctuations. However, in cases where a project has a life longer than one day, usually between 2 and 10 years in the manufacturing world, it is critical to take cost fluctuations into consideration. It is also important to understand the project cash outflows and inflows since these determine the ROI of a project.

How do costs fluctuate over time? Although some might argue that trying to predict the economic future is futile, they are actually predicting it by not trying, because they are in effect predicting that the future will not change. By assuming today's cost, they assume that this cost will not change over the life of a project. This assumption is incorrect, because life changes and thus cost becomes a moving target.

For example, labor cost in China has been steadily increasing throughout the 1990s and 2000s, thus it is probably safe to assume that the labor cost will continue to go up in the following 5 to 10 years. Similarly, electricity and gas prices have been trending up as global demand goes up and supply decreases. For the same reason, raw material prices for steel, iron, aluminum, copper, precious metals, and oil all have been increasing. A good cost estimator will be watching the trends very closely and will have to play a role of an amateur economist to account for those future trends in his/her estimates.

This is done using business case modeling. Table 2.4 shows a business case example that simulates cost fluctuations over the program life. It also includes internal rate of return (IRR) and payback calculations that help cost estimators evaluate project worthiness in addition to simple profit assessment.

This business case estimate leads to yet another price: $3 per connector. That is because the business case now considers increases (inflation) in future raw material (3%), labor (3%), and manufacturing overhead (3% due to electricity, gas, and

TABLE 2.4

Sample Business Case Calculation with One-Year Payback

		Supplier Business Case						
		Year 1	Year 2	Year 3	Year 4	Year 5		
	Volume ('000)	10	10	10	10	10		
3%	Raw material	$0.04	$0.04	$0.04	$0.04	$0.04		
–2%	Purchased parts	$0.42	$0.41	$0.40	$0.40	$0.39		
	Materials, others	$0.18	$0.18	$0.18	$0.18	$0.18		
3%	DL	$0.04	$0.04	$0.05	$0.05	$0.05		
3%	Mfg. OH & others	$0.28	$0.29	$0.30	$0.31	$0.31		
	New investment	$0.20	$0.20	$0.20	$0.20	$0.20	$10.00	Total ('000)
	Cost of capital	$0.09	$0.09	$0.09	$0.09	$0.09	$4.50	Total ('000)
2%	Scrap	$0.02	$0.02	$0.03	$0.03	$0.03		
	Total mfg. cost	$1.27	$1.27	$1.28	$1.28	$1.28		
	SGA	$0.50	$0.50	$0.50	$0.50	$0.50		
	SGA ('000)	$5.00	$5.00	$5.00	$5.00	$5.00	$25.00	Total ('000)
	% SGA	16.7%	17.2%	17.7%	18.3%	18.8%	17.7%	Lifetime
	Packaging	$0.05	$0.05	$0.05	$0.05	$0.05		
	Ship/others	$0.03	$0.03	$0.03	$0.03	$0.03		
	Profit	$1.15	$1.06	$0.97	$0.88	$0.79		
	% Profit	38.3%	36.3%	34.3%	32.1%	29.9%	34.3%	Lifetime
	Profit ('000)	$11.50	$10.58	$9.67	$8.79	$7.93	$48.47	Total ('000)
–3%	Price	$3.00	$2.91	$2.82	$2.74	$2.66		
	Revenue ('000)	$30.00	$29.10	$28.23	$27.38	$26.56	$141.27	Total ('000)
Year	0	1	2	3	4	5		
Cash flow	($10.00)	$10.05	$9.40	$8.77	$8.15	$7.55		
IRR	91%							
Payback	1.00	Years						
10	Years of depreciation							
9%	Borrowing interest rate							
$4.50	Total ('000)							
30%	Tax rate							

indirect labor cost increases). Additionally, the supplier expects the customer to demand annual price decreases of 3% per year (sometimes called productivity give-backs). Finally, the one-year payback requirement, or hurdle rate, drives a higher price because of the necessary $10,000 investment needed for this specific project. This is the additional dimension that cash flow consideration brings to the price calculation.

The supplier can make an informed decision to forgo the one-year payback requirement and still have a very good IRR. For example, per Table 2.5, by changing the requirement to a two-year payback, the price can be reduced to $2.32 per part while reducing the IRR from 91% to still robust 36%.

There could also be a case where the investors do not care as much about payback, but are more interested in a certain IRR hurdle rate. Perhaps they feel that every

TABLE 2.5
Business Case Calculation with Two-Year Payback

		Supplier Business Case						
		Year 1	Year 2	Year 3	Year 4	Year 5		
	Volume ('000)	10	10	10	10	10		
3%	Raw material	$0.04	$0.04	$0.04	$0.04	$0.04		
−2%	Purchased parts	$0.42	$0.41	$0.40	$0.40	$0.39		
	Materials, others	$0.18	$0.18	$0.18	$0.18	$0.18		
3%	DL	$0.04	$0.04	$0.05	$0.05	$0.05		
3%	Mfg. OH & others	$0.28	$0.29	$0.30	$0.31	$0.31		
	New investment	$0.20	$0.20	$0.20	$0.20	$0.20	$10.00	Total ('000)
	Cost of capital	$0.09	$0.09	$0.09	$0.09	$0.09	$4.50	Total ('000)
2%	Scrap	$0.02	$0.02	$0.03	$0.03	$0.03		
	Total mfg. cost	$1.27	$1.27	$1.28	$1.28	$1.28		
	SGA	$0.50	$0.50	$0.50	$0.50	$0.50		
	SGA ('000)	$5.00	$5.00	$5.00	$5.00	$5.00	$25.00	Total ('000)
	% SGA	21.6%	22.2%	22.9%	23.6%	24.3%	22.9%	Lifetime
	Packaging	$0.05	$0.05	$0.05	$0.05	$0.05		
	Ship/others	$0.03	$0.03	$0.03	$0.03	$0.03		
	Profit	$0.47	$0.40	$0.33	$0.26	$0.19		
	% Profit	20.3%	17.7%	15.0%	12.2%	9.3%	15.1%	Lifetime
	Profit ('000)	$4.70	$3.98	$3.27	$2.59	$1.91	$16.45	Total ('000)
−3%	Price	$2.32	$2.25	$2.18	$2.12	$2.05		
	Revenue ('000)	$23.20	$22.50	$21.83	$21.17	$20.54	$109.25	Total ('000)

Year	0	1	2	3	4	5
Cash flow	($10.00)	$5.29	$4.79	$4.29	$3.81	$3.34
IRR	36%					
Payback	2.00	Years				

10	Years of depreciation
9%	Borrowing interest rate
$4.50	Total ('000)
30%	Tax rate

project must have at least 20% IRR, otherwise they can invest their money elsewhere. In this case, the supplier could lower its part price even further to $2.15, but also increase the payback to three years, as shown in Table 2.6.

Regardless of which calculation the supplier chooses, the market will determine the actual price of any product, including our connector. What the calculation

TABLE 2.6
Business Case Calculation with 20% IRR

		Supplier Business Case						
		Year 1	Year 2	Year 3	Year 4	Year 5		
	Volume ('000)	10	10	10	10	10		
3%	Raw material	$0.04	$0.04	$0.04	$0.04	$0.04		
−2%	Purchased parts	$0.42	$0.41	$0.40	$0.40	$0.39		
	Materials, others	$0.18	$0.18	$0.18	$0.18	$0.18		
3%	DL	$0.04	$0.04	$0.05	$0.05	$0.05		
3%	Mfg. OH & others	$0.28	$0.29	$0.30	$0.31	$0.31		
	New investment	$0.20	$0.20	$0.20	$0.20	$0.20	$10.00	Total ('000)
	Cost of capital	$0.09	$0.09	$0.09	$0.09	$0.09	$4.50	Total ('000)
2%	Scrap	$0.02	$0.02	$0.03	$0.03	$0.03		
	Total mfg. cost	$1.27	$1.27	$1.28	$1.28	$1.28		
	SGA	$0.50	$0.50	$0.50	$0.50	$0.50		
	SGA ('000)	$5.00	$5.00	$5.00	$5.00	$5.00	$25.00	Total ('000)
	% SGA	23.3%	24.0%	24.7%	25.5%	26.3%	24.7%	Lifetime
	Packaging	$0.05	$0.05	$0.05	$0.05	$0.05		
	Ship/others	$0.03	$0.03	$0.03	$0.03	$0.03		
	Profit	$0.30	$0.23	$0.17	$0.10	$0.04		
	% Profit	14.0%	11.2%	8.3%	5.3%	2.1%	8.3%	Lifetime
	Profit ('000)	$3.00	$2.33	$1.68	$1.03	$0.41	$8.45	Total ('000)
−3%	Price	$2.15	$2.09	$2.02	$1.96	$1.90		
	Revenue ('000)	$21.50	$20.86	$20.23	$19.62	$19.03	$101.24	Total ('000)

Year	0	1	2	3	4	5
Cash flow	($10.00)	$4.10	$3.63	$3.17	$2.72	$2.28
IRR	20%					
Payback	3.00	Years				

10	Years of depreciation
9%	Borrowing interest rate
$4.50	Total ('000)
30%	Tax rate

provides is information that helps management to make a better decision of whether the business is worth pursuing. For example, if the customer's target price is around $1.03 and there is another supplier that quotes the project at this level, then our connector supplier might just decide to walk away. Or, it can make a strategic decision to lose money on this project in order to appease the customer and win bigger and more profitable projects in the future.

Conversely, the customer would benefit from having a transparent relationship with its suppliers. By knowing that it is sourcing unprofitable projects to its suppliers, it can manage the risk of supplier bankruptcies or project failures and thus maintain a better flow of parts to its factories.

ANNUAL PRODUCTIVITY GIVEBACK DISTORTION

It was briefly mentioned earlier, and requires further explanation, that the customer may require its suppliers to reduce prices annually. This is common in some industries (e.g., automotive) where customers expect suppliers to continuously improve pricing by achieving higher productivity of their manufacturing processes and supply chain. It is assumed that through cycle time improvements, operator reductions, more localized supply base, and various other cost reduction opportunities, the suppliers will reduce their cost over time.

However, with the inflationary pressures on raw materials, labor, and energy, in addition to difficulty in changing manufacturing processes and supply chain after production start due to requirement for customer approvals, the companies supplying to customers that require annual price reductions have had to adjust by adding future reductions to their first-year price. In those instances, the prices are automatically higher than they should be (see Figure 2.7). Unfortunately, even though this is common knowledge, most customers have not adjusted their practices. This is because of the pressure by executive management to show annual cost structure reductions, which then results in their procurement departments having the performance objectives based on those annual price reductions.

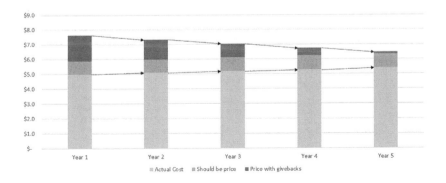

FIGURE 2.7   Price impact of givebacks.

LUMP SUM PAYMENT DISTORTION

Another common practice that distorts pricing is the requirement by some custom-ers to have suppliers provide upfront lump sum payment in order to secure future business. However, if suppliers have not accounted for cash outflows in their cur-rent pricing or have no current business with this customer, they will simply have to resort to "baking in" this payment into their future pricing. As Figure 2.8 shows, this effectively ends up being a high interest loan to the customer (supplier investors typically have higher expectations for return than the open financial market).

What the customers using the practice of productivity givebacks and lump sum payments might find is that their overall cost structure is not really decreasing over time despite the annual cost savings they achieved (see Figure 2.9). This is due to the fact that the money received from suppliers annually is offset by higher prices on new projects launched that year.

It is recommended that these practices are abolished and that parts are sourced at best price upfront and without a requirement for annual productivity givebacks and upfront lump sum payments. However, this should not prevent the customer from using new business leverage to achieve lower pricing. In the long term, the use of cost

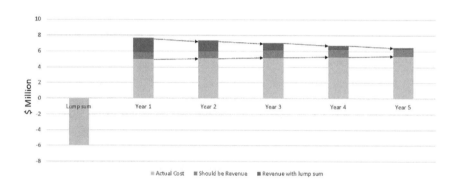

FIGURE 2.8 Price impact due to lump sum payments.

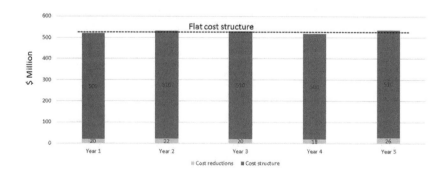

FIGURE 2.9 Cost structure impact of distortions.

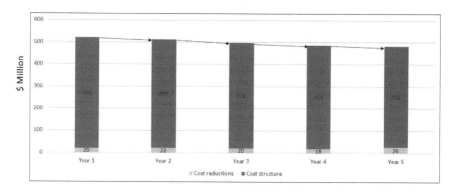

FIGURE 2.10   Cost structure impact with cost engineering.

engineering methodologies with real cost optimization efforts should bring on real improvement in cost as seen in Figure 2.10.

QUOTING PROCESS

As a supplier quoting business to customers, there are a couple of key ingredients that are needed to be successful in winning profitable business. First, since quoting is a cross-functional team sport, the objectives must be aligned between participants in the quoting process. For example, the sales representative's objectives are often centered on revenue only and do not consider profit. In that situation, the sales rep will often focus simply on winning the business regardless of the profit, which could be detrimental to the success of a quoting process. Another example is that of "sand bagging," where participants from purchasing or manufacturing provide higher than necessary cost to mitigate any future risks.

This risk aversion type of behavior could significantly increase quoted prices and severely affect chances of winning business. It is therefore critical that the quote process participants are aligned on defining real and reasonable cost while quoting business at satisfactory profit. As the team goes through many rounds of quoting for the same business, they must resist the temptation to "sharpen their pencils," a direction often given by customers, which could drive the quote team to manipulate costs just to get to the winning price. The focus instead should be on defining the minimum acceptable or "walk away" price and then letting sales fight for the best possible price above that.

Second, the quoting process must be timely. It is important to keep in mind that the quoting process should be based on customer's time requirements, not on the time it takes a supplier to put together a quote. If a customer typically only gives two weeks for a quote response, then this is the maximum time it should take the supplier to send out a quote. Unfortunately, this is not always that simple mainly because there could be several people involved from various functions, and they all need to perform certain tasks that each takes time.

To achieve quote timing requirements, it is advisable to define a very specific process. This process must give engineering time to complete the design with sufficient detail, give manufacturing time to define a process and investment needed for that design, give sub-suppliers enough time to quote various parts (or give cost estimating enough time to simulate sub-supplier pricing using cost models), and also allow enough time for finance and sales to put it all together, analyze the business case, and present it to the executive management.

To manage all these pieces of the quoting process, a quoting manager should be assigned. This person will keep the team on track, consolidate data, develop the business cases, and hold firm to maintaining profit integrity for the company. It is also useful to employ software that serves as a data hub for the team providing each member with data transparency and documentation. This is most often done using MS Excel, but there is also specialized software available that allows for better tracking throughout the quoting process and product life.

CHANGE MANAGEMENT CONTROL

An often-overlooked aspect of cost control is the change management part of it. After the business is awarded, the cross-functional team moves on to the next project, yet the design and manufacturing process for the awarded business will most likely continue to go through multiple rounds of changes. Each of the changes must be understood from the cost perspective and most likely reported to the customer. This requires a disciplined approach to evaluating, tracking, and reporting of the cost impacts. It is helpful to have an ERP or a PLM system that gathers inputs from all involved parties needed in the evaluation process and which gives access to all those that need to be aware of the outputs. For significant changes, the quoting process might have to be employed to provide a well-thought-out analysis and pricing to the executive management and to the customer. Without a disciplined approach to change management control, companies often find themselves in a position where a business that was originally awarded at required profit suddenly deteriorates into loss and no time to improve the situation with very little sympathy from the customer.

SUMMARY

Cost estimating capability is critical in the cost engineering process. Much like a navigation tool guides a navigator, cost estimates guide the organization toward its cost targets. Understanding the basic cost categories – such as raw material, purchased components, packaging, logistics, labor, overhead, depreciation, setup, scrap, and tooling – is key in the decision making process. They provide those working in manufacturing the necessary knowledge to navigate financial waters.

It is important to keep in mind that there are various degrees of cost estimating complexity. The most common estimates are based on point in time assumptions, meaning estimating what things cost today. A more complex, but probably more accurate, type of cost estimating is based on defining the full business case over a program or product life. These type of estimates require more foresight into the future and assumptions about what might happen to various costs due to regional and global economic factors.

The business case type of cost estimating might also involve some assumptions about customer and supplier behaviors, mainly the requirements to reduce customer price over time and the company's ability to match those price reductions with same or similar price reduction of purchased materials and components. In those cases, it is critical to understand the real impacts on the bottom line of the company over time and to avoid any distortions due to givebacks and long term agreements.

Quoting business is very complex and requires the supplier to have an organizational alignment on objectives and built in timeliness to the quoting process. The focus should be on defining real and reasonable cost while maintaining profitability for each business case. The quoting process should be defined in such a way that it allows time for each function to complete its task. A quoting manager should be assigned to the quoting process and tools, such as software, should be employed to help the team with data collection and documentation.

3 Cost-Based Negotiation

Unless a company has a completely open book collaborative relationship with its suppliers or customers, there is bound to be a dispute over what the product price should be. Business's main purpose is to make as much profit as possible. It is not unreasonable that both customers and suppliers will do everything they can to justify their pricing assumptions, regardless of whether they reflect real cost or not. In some cases, the low levels of collaboration are akin to suppliers feeling like they are being chased by customers with price hammers (see Figure 3.1). As the collaboration grows, it can still feel like a high-stakes poker game with both sides holding cards close to their chests.

It is only for those companies that have a collaborative, open book relationship that the negotiation will be based purely on cost data and its underlying assumptions without the uncomfortable lying. That is very different than a typical negotiation which is based on price data points only, meaning that without any knowledge of cost behind the pricing, buyers will simply focus on collecting competing quotes, then choose the lowest price (this is sometimes called the "Collect & Compare" technique). The negotiation, in a typical process, is simply leveraging suppliers against each other to the lowest point, regardless of what the actual cost might be.

In the cost engineering process, the focus is on understanding the cost assumptions and resolving any gaps between cost estimates from each side with the goal of finding cost reduction opportunities that will ultimately lead to lower prices (see the nine-step process in Figure 3.2). Even if the initial price is higher than other competitors, the quote would not be automatically disqualified, but instead, a process would begin to understand it and find ways to improve it. Note that the key is a mandatory and standardized cost breakdown submission by the supplier. This is required for the gap closure process to be successful; nothing can be achieved without it. Online cost breakdown collection software can make this requirement less painful for both suppliers and customers.

For companies with a cost engineering culture already in place, the collaboration with suppliers would be already at a high level, so this fact-finding process would only be needed with one or two trusted suppliers. This is because early cost engineering engagement means that time will only permit engagement with one or two suppliers; any more than that would be time and resource prohibitive.

Let's use our connector as an example and let's assume that the supplier and the customer have a typical non-collaborative relationship without cost engineering focus. Both the supplier and the customer cost estimators have come up with estimates, and there is a very large gap between the two estimates, $2.5711 and $1.0025, respectively. Not certain about the accuracy of its cost estimate assumptions, the

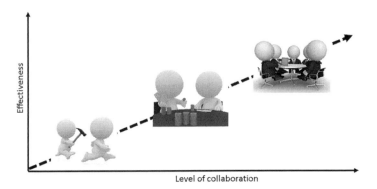

FIGURE 3.1   Collaboration vs. effectiveness chart. (Courtesy of freestock.com; https://www.freestock.com/free-photos/3d-guy-chasing-hammer-isolated-white-59466709, https://www.freestock.com/free-photos/3d-people-playing-cards-gambling-isolated-84196423, https://www.freestock.com/free-photos/3d-business-people-corporate-meeting-isolated-73351390.)

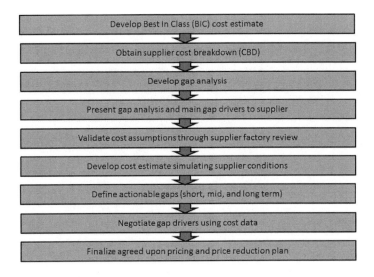

FIGURE 3.2   Nine-step gap closure process.

supplier's sales person only provides the price of $2.5711 to the customer without giving any assumptions behind it. The customer's buyer already has an estimate of $1.00 from his/her cost estimator, so he/she gives feedback to the supplier that their estimate of $2.5711 is "way off" and to "sharpen their pencils," then to provide a cost breakdown for evaluation.

The supplier's quoting team goes back to the drawing table and challenges every assumption that they have made. Unfortunately, after much effort, the team decides that all the assumptions are correct. They believe that two specific cost buckets are probably the reason why the price ended up higher than what the customer expected, corporate overhead at $0.50 and profit at $1.00 per piece. These two ended up higher because of the low volume that is in turn driving low cash flow, which is not sufficient to overcome the needed investment and expenses for this project.

The supplier does not feel that their customer would be willing to accept the real reasons for this higher than expected price and would probably only accept a flat 10% and 5% for corporate overhead and profit, respectively. The supplier is faced with two options: refuse to provide a cost breakdown or provide a cost breakdown but one where the high corporate overhead and profit are hidden in other cost buckets, such as manufacturing overhead, depreciation, or purchased parts. The customer would have a more difficult time challenging those cost buckets.

Faced with this situation, many suppliers choose not to provide a cost breakdown, maybe even going as far as suggesting that this is their corporate policy. If the customer does not require cost breakdowns, then this supplier's quote will perhaps be simply rejected and another supplier will be chosen, perhaps one that does not understand cost, or one that is willing to take this business at a loss in hopes of positioning itself for a profitable high-volume business in the future.

In our case, the supplier decides to go with the second option and provide a modified cost breakdown with Profit and SG&A hidden in other cost buckets (purchased parts and manufacturing overhead), as seen in Table 3.1 below. This is to get a chance at continuing the sourcing process with hopes of selling the customer on the price or on other value that the supplier provides, such as technology or quality. The customer may still decide to go with a lower bid, if one exists, but at least there is a chance that the customer will analyze the cost breakdown and continue the dialogue.

The customer does not provide feedback on this updated cost breakdown, and after some time the supplier finds out that the business was sourced to another supplier. In this typical non-collaborative environment, the buyer simply skipped the cost analysis and sourced the lowest of the supplier quotes or sourced the supplier that the buyer feels more comfortable with for whatever reasons.

The supplier feels disenchanted, but there is a good chance that both supplier and customer missed an opportunity in this scenario. The supplier lost potential business, but the customer could have negotiated a lower price with our connector supplier only if some negotiation took place on the assumptions of each cost bucket. The customer is also taking on a risk because it might have sourced a supplier that gave a lower price due to lack of understanding of its own cost which may lead to problems down the road (e.g., supplier financial issues). There is a very high possibility that this typical sourcing process will lead to inefficiency of pricing and ultimately lower profits for the customer and/or supplier.

How would this sourcing process work if supplier and customer were in a collaborative, open book relationship? First, the supplier would not be afraid to show its real cost to the customer. In fact, the customer will already know what the supplier's cost would be since the customer's cost estimator would have a pre-agreed cost model simulating the supplier's cost structure. This in itself would eliminate

TABLE 3.1

Modified Cost Breakdown Comparison

	Supplier Cost Breakdown to Customer	Supplier Assumptions	Customer Cost Estimate	Customer Assumptions
Raw material	$0.0375	15 grams/part, $2.50/kg	$0.0260	13 grams/part, $2.00/kg
Purchased parts	$0.7984	Sub-supplier quotes	$0.4800	Historical data
Packaging	$0.0042	1% of material cost	$0.0048	1% of material cost
Inbound logistics	$0.0210	Logistics quote	$0.0101	2% of material cost
Material overhead	$0.1500	Historical data	$0.0506	10% of material cost
Direct labor	$0.0430	15 sec/part, 2 ops, 85% efficiency	$0.0268	13 sec/part, 1.5 ops, 90% efficiency
Indirect labor	$0.0167	$48 per hour	$0.0134	50% of direct labor cost
Setup	$0.0950	$19 per setup, 200 parts per run	$0.0019	$19 per setup, 10,000 parts per run
Manufacturing overhead	$0.8922	$150 per hour	$0.1453	Assy.: $5/hr, mold: $30/hr
Depreciation	$0.0166	Assy.: $0.023/hr, mold: $2.30/hr	$0.0211	Assy.: $0.25/hr, mold: $4/hr
Scrap	$0.0415	2% of incurred cost	$0.0156	2% of incurred cost
Total Manufacturing Cost	*$2.1161*		*$0.7956*	
Corporate overhead (SG&A)	$0.2500	10% markup	$0.0796	10% markup
Profit	$0.1250	5% markup	$0.0796	10% markup
Price	*$2.4911*		*$0.9547*	
Outbound packaging	$0.0500	$100 per 1,000 parts per pallet	$0.0191	2% of price
Outbound logistics	$0.0300	Shipping supplier quote	$0.0286	3% of price
Price w/Packaging and Logistics	*$2.5711*		*$1.0025*	

many disconnects between the two estimates. The customer's cost estimator could be using the pre-agreed cost model to estimate costs without ever requiring quotes from the supplier. This would be particularly useful during the product development phase where cost guidance is needed without much time. Second, the pre-agreed cost models give both sides security that the estimates are accurate enough that neither supplier nor customer will have significant financial risk. The customer can accurately predict their cost without risk of supplier financial issues while the supplier can secure business that is profitable.

TABLE 3.2
Cost Breakdown Comparison (Unmodified)

	Supplier Cost Breakdown to Customer	Supplier Assumptions	Customer Cost Estimate	Customer Assumptions
Raw material	$0.0375	15 grams/part, $2.50/kg	$0.0260	13 grams/part, $2.00/kg
Purchased parts	$0.4200	Sub-supplier quotes	$0.4800	Historical data
Packaging	$0.0042	1% of material cost	$0.0048	1% of material cost
Inbound logistics	$0.0210	Logistics quote	$0.0101	2% of material cost
Material overhead	$0.1500	Historical data	$0.0506	10% of material cost
Direct labor	$0.0430	15 sec/part, 2 ops, 85% efficiency	$0.0268	13 sec/part, 1.5 ops, 90% efficiency
Indirect labor	$0.0167	$48 per hour	$0.0134	50% of direct labor cost
Setup	$0.0950	$19 per setup, 200 parts per run	$0.0019	$19 per setup, 10,000 parts per run
Manufacturing overhead	$0.1677	Assy.: $15/hr, mold: $20/hr	$0.1453	Assy.: $5/hr, mold: $30/hr
Depreciation	$0.0166	Assy.: $0.023/hr, mold: $2.30/hr	$0.0211	Assy.: $0.25/hr, mold: $4/hr
Scrap	$0.0194	2% of incurred cost	$0.0156	2% of incurred cost
Total Manufacturing Cost	*$0.9911*		*$0.7956*	
Corporate overhead (SG&A)	$0.5000	$5,000 cost per year	$0.0796	10% markup
Profit	$1.0000	1-year payback on $10k investment	$0.0796	10% markup
Price	*$2.4911*		*$0.9547*	
Outbound packaging	$0.0500	$100 per 1,000 parts per pallet	$0.0191	2% of price
Outbound logistics	$0.0300	Shipping supplier quote	$0.0286	3% of price
Price w/Packaging and Logistics	*$2.5711*		*$1.0025*	

For the sake of argument, let's assume that in the case of our connector, the preagreed cost model does not work well for low-volume projects and there is, indeed, a large gap of $1.5687 between supplier and customer cost estimates, shown again in Table 3.2.

It is often helpful to both customer and supplier to also have a more graphical representation of the gap like the waterfall chart in Figure 3.3. This clearly identifies gaps between cost estimates and gives a Pareto analysis of which gaps are most impactful.

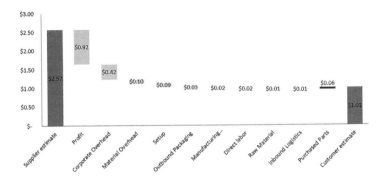

FIGURE 3.3   Gap analysis in graphical form.

Since there is a collaborative open book relationship between supplier and customer, the cost breakdowns are transparent to both sides and comparative analysis can begin. Below is what a cost-based negotiation might look like for each cost bucket.

RAW MATERIAL

For this cost bucket, the costs are $0.0375 and $0.0260 per supplier and customer cost estimates, respectively. After comparing the assumptions, it is found that the reasons for this discrepancy are that the supplier assumed higher part gross weight (15 vs. 13 grams) and higher resin cost per kilogram ($2.50 vs. $2.00).

Both sides feel strongly about their assumptions, so what ensues now is really a data-driven or fact-based negotiation. The goal is to convince the other side to change their assumptions or perhaps be convinced, with evidence, to change their own. The supplier might claim that their preliminary resin flow analysis recommends 30 grams as a correct shot weight for two-cavity mold. However, since the supplier might tend to be conservative in initial estimates, the customer's cost estimator or buyer could perhaps argue that based on similar parts currently produced, 26 grams is more accurate.

If the customer's cost estimator is experienced enough, he/she might be able to challenge the initial mold layout proposed by the supplier. Ultimately, the two sides might agree on a shot weight that is between 30 and 26 grams or the customer might task the supplier to work toward designing a lower shot weight prior to production.

The other component of the discrepancy is the resin cost. The supplier might argue that a particular resin that the customer requests is not widely used and that they do not have much leverage with a particular resin supplier for such low annual quantity. The customer, not having data for lower resin order quantities, might agree with the supplier and instead task its own engineering group to change the resin to a more commonly used one. Or, the customer might decide to challenge the supplier to go back to its resin supplier and improve their quote by leveraging other resin purchases.

It is worth noting that the transparency and open discussion already resulted in two cost reduction ideas, improving the mold layout to reduce shot weight and changing to a more common resin. These could potentially result in price reductions that would not have been otherwise identified.

PURCHASED COMPONENTS

In this case, the supplier's cost is actually lower than what the customer assumed, $0.42 versus $0.48. Most likely, the team will move ahead with the lower number and focus on other items, but this might be a missed opportunity. Perhaps the team should review the sub-supplier assumptions to check if any cost reduction opportunities exist. The supplier should have the same collaborative approach with its sub-suppliers as it has with its customer. The cost engineering philosophy and process should be extended across the full value chain. This would be especially true if a customer did not have an open book collaborative relationship with its suppliers and, as in our modified cost breakdown example, the suppliers tried to hide excessive profit or cost in the purchased components.

PACKAGING (INBOUND)

This is a very small part of the overall cost and there is only a very small discrepancy, $0.0048 versus $0.0042. Since both sides assume that packaging is 1% of material cost, the gap will be closed if agreement is reached on what the material cost should be. Or, an actual packaging cost can be developed based on packaging requirements.

LOGISTICS (INBOUND)

Although the gap for this cost bucket is also very low, $0.0210 versus $0.0101, it might be worth exploring why the supplier feels that approximately 4% of material cost is more appropriate than 2% assumed by the customer. Does the supplier assume that the resin price excludes delivery? From where is the supplier shipping purchased components? Answering these and other questions might lead to further opportunities, even if they are only fractions of a penny.

MATERIAL OVERHEAD

In this case, the supplier and customer cost estimators used completely different assumptions for their estimates. The supplier cost estimator used historical data indicating that it will cost the supplier $1,500 per year to manage the material (procurement buyers, material planners, warehouse and its staff, forklift operators, etc.). On the other hand, the customer cost estimator thought this part will require higher than usual material overhead cost, but assigned only 10% of material cost. This resulted in a large cost discrepancy, $0.1500 versus $0.0506 per piece, or $1,500 versus $506 per year.

It is extremely difficult for either side to argue for the accuracy of their assumptions because it is usually very difficult to allocate material overhead cost to specific products. Only if proper drivers are assigned to each cost activity can a fairly accurate cost be allocated. In the customer's case, assigning this cost by percentage is the easiest way to allocate cost, but is rarely accurate. It would be difficult for the customer to argue that $506 per year is enough or that 10% on average is applicable. On the other hand, if the supplier has good information behind their $1,500 per year assumption, then it probably has an upper hand in the negotiation. Perhaps the supplier is using activity-based costing and can allocate costs based on order quantities (e.g., it costs $500 per pallet ordered per month) or by invoice processed (e.g., it costs $100 per invoice). That would make it easier for the supplier to justify their cost and maybe even educate their customer and force them to update their cost model.

DIRECT LABOR

The gaps in direct labor cost are usually easy to resolve because it is the most transparent cost component. A person can simply walk out onto the manufacturing floor and see how many operators are running the machines and how long it takes to make a single part, the two critical pieces of information needed for the direct labor cost calculation. However, there are some tricks that suppliers might play if you don't have a collaborative open book relationship with them, such as putting extra operators at the machine or deliberately slowing down the machine whenever the customer visits. Nevertheless, these tricks are not advisable and can easily backfire if the customer has an experienced buyer or cost estimator who has seen many similar parts produced at various suppliers.

For our connector, there is only a slight discrepancy between supplier and customer estimates, $0.0430 versus $0.0268, respectively. The gap between the two is mostly driven by the cycle time (time to make a single part or time between parts produced at the end of the line) and slight differences in number of operators, 2 versus 1.5, and efficiency, 85% versus 90%. The customer assumed that the cycle time will be 13 seconds per part for both assembly and over-molding operations; however, the supplier feels that assembly will actually take about 15 seconds. The suppliers usually perform "motion study" analysis to determine, motion by motion, how long it will take an operator to perform all the needed movements. Since it is common for manufacturing engineers to be conservative with initial process assumptions (mostly driven by tough post production efficiency improvement targets), the team should review the motion study to make sure it is not too conservative. It is important that both customer and supplier challenge the assumptions in order to arrive at the most realistic assumptions so that the production price already has all the waste taken out of it.

After reviewing the motion study, the team agreed that 15 seconds is probably the correct production cycle time assumption for the assembly. Both sides also agree that 13 seconds is reasonable for the over-molding process. It is important to note here that having two different cycle times is acceptable in this case only because the assembly and over-molding processes are not sequenced, meaning that one occurs separately from the other and they are not dependent on each other. They occur on separate machines in different departments within the factory.

If the machines were sequenced, meaning that one machine was waiting for the other to finish in order to make a part, then the slowest cycle time would have to

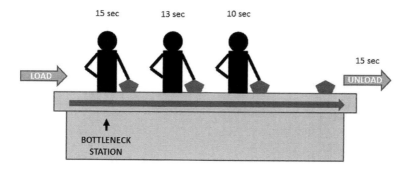

FIGURE 3.4   Bottleneck constraint in the manufacturing process.

be assumed for both machines/processes. Since one machine would be waiting for the other to finish, the operator and machine would be "captive" of the other process, and the cost for both would occur even if no parts are being made. Figure 3.4 is a graphical representation of this concept, also known as theory of constraints, explained very well by Eliyahu M. Goldratt in his book "The Goal: A Process of Ongoing Improvement."

The other two discrepancies, number of operators and efficiency, also need to be challenged. The supplier assumes an operator for each process, but maybe one operator can operate two molding presses since a robot will perform the insertion of an assembly into the press. Also, the customer can challenge the efficiency assumption based on market data for suppliers producing similar products.

It is worth noting that the labor rates can often lead to discrepancies as well. The customer might not have the labor rates for specific countries or regions within a country, so he/she must use assumptions based on publicly available data such as government websites. However, these usually only provide averages while actual industry and local wages can vary widely depending on labor supply at a specific location. For example, labor rates in Mexico can be lower than average if a factory is the only one in the specific town or area, but they can rise dramatically if other companies decide to open factories in that same area. The level of competition for labor, especially if skilled labor is needed and there is shortage of such labor in that specific location, can significantly impact the labor rates in any given region or city.

An important consideration is that it is often assumed that reductions in cycle time lead to a reduction in cost, but this is not always the case. If there are no operators actually being eliminated during any given work shift due to the cycle time reductions, then there will be no resulting cost reduction. For example, let's assume that the molding press is currently fully utilized producing five different products every day, 24 hours per day, six days per week, and 50 weeks per year. The only time the press is not running is during the planned breaks and lunches for the workers and during the unplanned machine breakdowns, which result in the 15% inefficiency that was assumed for the connector. Now, let's assume that the supplier is able to improve the cycle time for our connector from 15 seconds to 13 seconds. Does this allow for the elimination of a machine operator that runs the press on any of the shifts? The answer is probably not. Even with the lower cycle time, the operator still remains and has to be paid his/her hourly wage for the full shift. No real cost savings were achieved. If the supplier assumes that there are savings, it will simply be assuming lower profit for this part. Only long term, if the supplier drives continuous improvement across all its products, will it be able to eventually create enough open capacity to perhaps produce a sixth product on that press. If the press is dedicated to only one product and no other products can be produced there, then cycle time reductions will not produce any real cost savings.

INDIRECT LABOR

In the case of indirect labor, the customer's rough assumption of 50% of direct labor cost is actually very close to the supplier's cost estimate, which was based on allocating labor to specific machines and developing an hourly rate. This results in a minimal gap between $0.0134 and $0.0167 cost estimates. It is most likely that both the customer and supplier will agree to focus on other cost buckets in their negotiation.

In cases where there is a large gap in estimates, the negotiation might come down to allocating indirect laborers to specific machines or products. This can be done by factory reviews of existing allocations for existing products.

SETUP

Per Table 3.2, there is a large gap of $0.0931 between supplier and customer setup cost estimates of $0.0950 and $0.0019, respectively. The entire gap is driven by the assumption of how often the part production will be set up and for how long it will run. The customer assumes that due to the low volume of this product, the supplier plant will set up the machine and run parts only once per year, thus it is dividing the $19 setup cost over annual production requirement of 10,000 parts. The supplier will then send the required parts weekly to the customer's plant. Ignored are additional inventory holding costs at the supplier for the parts that have to sit on the shelf and wait to be shipped in the months ahead.

The supplier, however, does not want to make 12-months-worth of parts only to have them sit on a shelf for many months. Just like the customer, it runs a lean factory and does not want to tie up cash in inventory. The supplier assumes that it will produce weekly and only the quantity of parts needed that week. This means only about 200 parts and not 10,000 per run.

So, how to resolve this discrepancy? First, the customer must acknowledge that there will be additional cost on small-volume projects, either the cost of setting up more often or the cost of holding excess inventory. Second, both supplier and customer must understand the cost a supplier would incur if it held the inventory for up to 12 months, instead of just one week, and determine if that cost is higher than the cost of setting up more often. This inventory holding cost calculation is a cash consideration because the supplier is spending money to make parts that it cannot sell for up to 12 months. It can either use its own cash or borrow money from a bank to finance this expense. Since companies are constantly spending cash and getting paid later, they typically borrow money on short-term loans to bridge the gap. This allows them to only pay interest at any given time and not have to hinder their cash situation. Spending their cash might jeopardize their solvency and perhaps drive them into bankruptcy, therefore our supplier will probably borrow money at a certain short-term loan interest rate, and the cost of that loan is what the supplier cost would be for that inventory. Assuming a 12-month loan at 5% interest rate and the cost of inventory of $9,911 (assumed all 10,000 parts at material and manufacturing process cost of $0.9911, but a lower number of components can be assumed since that inventory will be sold throughout the year), the interest payment would be $270.47 per year or $0.0270 per part. Combining this cost together with the cost of setting up once a year, the total cost per part is $0.0289, which is lower than the cost of setting up once per week of $0.0950 per part and thus should be considered as a cheaper option for the supplier. The supplier agrees that setting up once per year is cheaper, but adds the cost of financing the inventory (known as inventory holding cost).

MANUFACTURING OVERHEAD

Since the assembly and molding processes are not synchronized and both supplier and customer agree to 15- and 13-second cycle times for assembly and molding operations, respectively, the manufacturing overhead cost is updated by the supplier to $0.1549 and by the customer to $0.1693. This eliminates most of the gap between their estimates.

The above case would probably be the easiest manufacturing overhead negotiation that ever took place. Unfortunately, normally this cost bucket is the biggest cost discrepancy, especially if the supplier to customer relationship is not open book and the supplier tries to hide other cost or profit in manufacturing overhead, as was shown before in the modified cost breakdown example. If this is the case, then the customer usually has limited data to prove the supplier wrong, other than using competitor information. However, there is a technique that can be helpful to at least move the negotiation forward. In this technique, the customer cost estimator can reverse engineer the supplier's profit & loss (P&L) statement (per Table 3.3) to verify if the supplier's manufacturing cost per piece would add up to the overall factory's manufacturing overhead.

Per Table 3.3, the total factory cost is estimated to be $9,670,000, including $2,570,000 for manufacturing overhead (Plant Burden). This estimate is based on data collected during the supplier manufacturing plant visit, such as number of employees (direct, indirect, and salaried), square footage of the plant, and capital employed. These have cost drivers and corresponding assumptions for each cost category and perhaps for each country or region where the supply base is located. For example, the cost per laborer or per square footage would be different for each country or region.

TABLE 3.3
Reverse Engineered P&L

		$/hr	Fringes	Hours	Annual Cost
		2.5	1.25	2000	$7,500
			Driver	Cost per	Plant Annual P&L
Employees		–		–	–
Hourly employees	People		200	$7,500	$1,500,000
Salaried employees	People		40	$40,000	$1,600,000
Building					
Building rental	Sq. feet		200,000	$2.00	$400,000
Building insurance	Sq. feet		200,000	$0.20	$40,000
Building depreciation	Yrs		30	$0	$0
Building utilities	Sq. feet		200,000	$0.25	$50,000
Building maintenance	Sq. feet		200,000	$0.05	$10,000
Equipment					
Equipment depreciation	Yrs		10	$40,000,000	$4,000,000
Equipment utilities	Units		1,000,000	$0.50	$500,000
Equipment maintenance	Capital		40,000,000	0.50%	$200,000
Equipment supplies	Capital		40,000,000	0.20%	$80,000
Indirect materials (solder, oils, rags, etc.)	Capital		40,000,000	0.25%	$100,000
Tools					
Perishable tooling	Capital		40,000,000	0.50%	$200,000
Tool room (supplies for tool maintenance)	Capital		40,000,000	0.50%	$200,000
Customer unpaid tooling	Yrs		4	$1,000,000	$250,000
Other plant expenses					
Supplies	Sq. feet		200,000	$0.10	$20,000
Outside services	Capital		40,000,000	0.20%	$80,000
Phones, computers, office supplies, etc.	People		240	$1,000	$240,000
Travel & entertainment	People		240	$500	$120,000
Shipping supplies	Capital		40,000,000	0.20%	$80,000
Total annual plant cost					$9,670,000
Direct labor cost					$1,500,000
Salary labor cost					$1,600,000
Equipment depreciation					$4,000,000
Plant burden					$2,570,000
Ratios					
Overhead (w/o equipment depreciation) to direct labor					278%

TABLE 3.4
Overhead Rate Comparison

OH rate provided in breakdown	$150.00
Annual up time hours	4800
Hours per shift	8
Shifts per day	3
Days per week	5
Weeks per year	50
Efficiency	80%
Annual cost per year (w/depreciation)	$720,000
Annual depreciation	$100,000
Original cost of equipment	$1,000,000
Number of years	10
Annual cost per year (w/o depreciation)	$620,000
Plant capacity (or cost) utilized by product	5%
Projected plant cost with burden rate	$14,400,000
Actual plant cost	$9,670,000
Gap to actual plant cost	$4,730,000
	49%

The next step is to compare the rates that the supplier quoted to those derived from a reverse engineered P&L (per Table 3.4). The hourly burden rate quoted by the supplier ($150 per hour) results in total plant cost of $14,400,000, which is above the reverse engineered total of $9,670,000. Now, a data-driven negotiation can begin to further understand the numbers. Each assumption can be discussed and analyzed further until gaps between estimates are closed. Without any data, this negotiation would have become very difficult and would probably stall.

DEPRECIATION

Although the methods of calculation were completely different by the supplier and the customer cost estimators, there is actually very little gap between their estimates for depreciation ($0.0166 versus $0.0211, respectively). In fact, the customer's estimate is higher than the supplier's. Perhaps this is why it is often the case that suppliers agree to the typical customer calculation (original equipment cost divided over allowable years of depreciation per GAAP rules and over annual production hours), since this is the most conservative calculation method. Suppliers often employ used machines, and those machines might have much longer life than the GAAP allowable years of depreciation. Granted, older machines will require higher maintenance cost, which customers usually exclude from their calculation since they assume new machines, but unless the machines are very old, the maintenance cost should not be excessive. On the other hand, if the supplier employed very old machines, then it should probably be planning for significant future investments, which then would equate to assuming the customer's typical new machines method.

It is worth mentioning here the case of custom equipment, such as product specific assembly lines. In this case, a supplier will rarely have existing "sunk" equipment that can be used to manufacture the customer's product, and most likely the custom equipment will not be useful for other products. Therefore, the investment should be amortized over program life volumes. For example, a $2M investment for a product with a contracted five-year life and annual volumes of two million would result in a piece cost of $0.20. It would be too risky to assume program life longer than this because there is no guarantee of program life extensions and the equipment will not be applicable to any other products.

SCRAP

In the case of scrap, there is very little gap between supplier and customer cost estimates ($0.0194 versus $0.0156). This is because both assume 2% of incurred cost and both estimates of incurred cost are fairly similar. Under this circumstance, the negotiators would probably just move on to another cost bucket. However, scrap is often a large driver of cost gap (e.g., paint process, which will be discussed later in this book), and it must be treated with high focus. Large gaps in scrap on a percentage basis are actually very common. This is especially true for unique, custom, or new products where a supplier might not have full confidence of producing parts at a low scrap rate, or it may assume higher scrap rate at launch only and not at full production. These assumptions need to be reviewed to ensure that they are reasonable and not overly conservative.

Another assumption that needs to be clear is what is actually being scrapped. The supplier will often track only the overall cost of scrap as percentage of sales or revenue (which is different than assuming scrap as percentage of material cost or manufacturing cost), and it might assume the same percentage for all products that it quotes. However, both supplier and customer should drive a much more accurate method where scrap for each product is tracked separately and it considers cumulative effect of different materials scrapped at each manufacturing process. For example, a connector manufactured on an automated assembly line might have a lower scrap rate than one manufactured using manual labor. As a supplier, it is important to have accurate data in order to make better decisions. If it was known that automated assembly line scrap is 0.5% and manual assembly line scrap is 5%, then management could make a decision to automate all future assembly lines if investment is justified.

Additionally, for our connector, the scrapped material at the assembly line is different than at the over-molding press. While only wires and pins would be scrapped at the assembly line, the resin would also have to be considered at the over-molding press.

CORPORATE OVERHEAD

This is where there is a large gap of $0.4204 ($0.5000–$0.0796) between supplier and customer cost estimates. As mentioned before, this is because of the completely different approach to estimating. The customer cost estimator simply applied the same percent markup on cost that he/she would to any other product (high or low volume, high or low complexity, etc.) with an argument that this would eventually average out. Meaning that perhaps a specific percentage might not be enough to cover cost on some products, but it would be too much on others. Over time, the customer would argue, this would result in the supplier having sufficient funds to cover corporate overhead expenses for all products.

Perhaps this makes sense if a customer is sourcing all of its products to just one supplier and there is a good balance between small- and large-volume products. In reality, any customer would usually have multiple suppliers, in which case the low-volume products that are underpriced would tend to end up with a single supplier that is unknowingly taking on unprofitable projects. Projects like these are often called "the dogs," and customers will have no scruples unloading those projects to unsuspecting victims.

Fortunately, in the case of our connector, the customer and supplier are in a collaborative, open book relationship, thus the customer agrees with the supplier's minimum $5,000 corporate charge per project or $0.50 per piece for the connector. This is with an understanding, however, that the supplier will also consider real costs for large-volume projects where a flat percentage markup might result in over allocation.

For example, if the annual volume for the same connector was 10,000,000, instead of applying a flat 10% SG&A markup, which would result in $796,000 of cost annually, the supplier instead should be willing to reduce the percentage to perhaps 2%. This would result in a more realistic cost of ~$159,000 annually, still more than three times the minimum $50,000 per project that the supplier requires.

PROFIT

As in the case of corporate overhead, there is a large gap of $0.9204 ($1.0000–$0.0796) in profit between supplier and customer cost estimates. The reason is very similar to the gap in corporate overhead. Simply applying the same 10% to every project results in a highly inaccurate estimate. A better way to evaluate any project is to calculate its ROI or payback. For our connector, because the supplier requirement is to recoup a $10,000 investment on this project in one year, it must have a profit per piece of $1.0000 ($1.0000 × 10,000 = $10,000).

Since the supplier and customer are in a collaborative, open book relationship, the customer may argue that a two- or three-year payback is more fair or may accept the $1.0000 per piece profit assumption, with an understanding that other large-volume projects will be treated on the same basis. For example, if the volume for the same connector was 10,000,000 annually with an investment of $100,000, then a one-year payback would be achieved with $0.0100 cost per connector or only about 1.25% profit markup.

As long as the supplier and customer agree to this method for both small- and large-volume projects, there should be no issues on either side. It would actually result in a much less risky financial situation for both. The customer would get fair pricing on its purchased parts without worrying about the supplier losing money and possibly going bankrupt, thus putting supply of parts in jeopardy, while the supplier would be much more secure in its financial situation overall and for any individual project.

Keeping in mind that projects are often cancelled or moved for various reasons (e.g., lack of sales, quality issues, etc.), a supplier suddenly losing an overpriced large-volume project while retaining the underpriced project might put that supplier in dire straits. It is much less risky to have all projects properly priced.

SUMMARY OF CONNECTOR NEGOTIATION

After reviewing all the assumptions, the customer adjusts its estimate and agrees with the supplier on a price of $2.4722 (see Table 3.5). Although this price is much higher than the customer originally estimated, the cost assumptions are real and accurate.

TABLE 3.5
Final Cost Breakdown Comparison (After Negotiation)

CBD After Negotiation

	Supplier Cost Estimate	Supplier Assumptions	Customer Cost Estimate	Customer Assumptions
Raw material	$0.0315	14 grams/part, $2.25/kg	$0.0315	14 grams/part, $2.25/kg
Purchased parts	$0.4200	Sub-supplier quotes	$0.4200	Sub-supplier quotes
Packaging	$0.0042	1% of material cost	$0.0042	1% of material cost
Inbound logistics	$0.0210	Logistics quote	$0.0090	2% of material cost
Material overhead	$0.1500	Historical data	$0.1355	30% of material cost
Direct labor	$0.0309	15s assy., 13 mold, 1/0.5 ops, 85% eff.	$0.0296	15s assy., 13 mold, 1/0.5 ops, 90% eff.
Indirect labor	$0.0167	$48 per hour	$0.0148	50% of direct labor cost
Setup	$0.0289	$19/setup, 10,000/ run; 5% financing	$0.0289	$19/setup, 10,000/run; 5% financing
Manufacturing overhead	$0.1549	Assy.: $15/hr, mold: $20/hr	$0.1693	Assy.: $5/hr, mold: $30/hr
Depreciation	$0.0166	Assy.: $0.023/hr, mold: $2.30/hr	$0.0211	Assy.: $0.25/hr, mold: $4/hr
Scrap	$0.0175	2% of incurred cost	$0.0173	2% of incurred cost
Total Manufacturing Cost	*$0.8922*		*$0.8811*	
Corporate overhead (SG&A)	$0.5000	$5,000 cost per year	$0.5000	$5,000 cost per year
Profit	$1.0000	1-year payback on $10k investment	$1.0000	1-year payback on $10k investment
Price	*$2.3922*		*$2.3811*	
Outbound packaging	$0.0500	$100 per 1,000 parts per pallet	$0.0500	$100 per 1,000 parts per pallet
Outbound logistics	$0.0300	Shipping supplier quote	$0.0300	Shipping supplier quote
Price w/ Packaging and Logistics	*$2.4722*		*$2.4611*	

Going forward, the customer cost estimators should revise their cost models and estimates to account for low-volume scenarios and pre-agree with suppliers to these cost models so that they both can estimate products in the same way. This would make their estimates more useful early in the product development process since they would be more predictive of the actual cost in production. In fact, this should have been the case already in a well-established open book collaborative relationship between customer and supplier. The large gaps in our connector estimates were for demonstrative purposes only. If the SPARK hybrid truck company used correct cost estimating principles on their purchased components early in the development, they would not have to wait until production launch to find out that low-volume consideration drives cost significantly higher.

TYPES OF COST ESTIMATES

It is worth mentioning that a company might choose to use cost estimate types other than the two basic ones, point in time and business case. It might also find it useful to have estimates that define best in class (BIC) scenario, supplier's specific condition, regional conditions, or a mixture of all of them. This variety will give a cost estimator better understanding of cost while negotiating with suppliers.

For example, it could be that the BIC manufacturing process for a particular part is fully automated for suppliers in high labor cost countries such as Germany or Austria. However, the supplier used by a customer is located in Mexico and uses a manual manufacturing process while most other suppliers in Mexico are using a semiautomated process. In this situation, a cost estimator might want to estimate the cost to make a product under all three scenarios: manual, automated, and semiautomated process in Mexico. This would give the full understanding of cost options.

ADDITIONAL NEGOTIATION TIPS

For those companies who are somewhere in between an open and a closed book relationship with their suppliers or customers, the summary in Table 3.6 might be helpful in their negotiation. Additionally, since any negotiation should be treated as a search for cost reduction opportunities, the table also contains potential opportunities that could be derived.

NEGOTIATION PREPARATION

Even though buyers normally negotiate with suppliers, the customer's cost estimator is often the lead negotiator for his/her company in the cost-based negotiation. The cost estimator should be working hand in hand with the buyer, but he/she will have a much higher level of cost expertise required for this type of negotiation.

TABLE 3.6
Negotiation Strategies and Potential Cost Reduction Opportunities

Type of Gap Driver	Negotiation Strategies	Potential Cost Reduction Opportunities
Material type	Should assume the same material per design specifications.	Use cheaper material. Relax material specification.
Material price	Request invoices.	Increase spending with supplier to get volume benefit.
	Provide supplier with market data or company internal data used in cost target.	Buy from different raw material supplier.
Gross material weight	Ask for weight calculation method. Have supplier provide samples of mold layout, strip layout, etc.	Increase number of cavities per mold. Improve mold layout. Reduce scrapped material (offal, gates, rises, etc.).
Scrap	Ask supplier to provide internal scrap data including Pareto analysis of root causes.	Provide specific steps to reduce scrap. Increase amount of material that is recycled.
Scrap resale value	Request invoices. Provide supplier with market data for scrap resale value.	Sell to different scrap buyer.
Material overhead	Request inventory and warehousing details.	Eliminate handling (i.e., warehousing, inspection, etc.).
	Make sure cost is allocated accurately between products.	Improve ordering frequency.
Packaging/logistics	Request packaging details (type, size, density, etc.). Request value stream map.	Change from expendable to returnable packaging or vice versa depending on business case. Source materials and parts locally.
Process flow/layout	Verify process with PPAP (production test run) information. Verify process with supplier quality. Obtain process flow and layout. Conduct supplier plant tour.	Optimize process to best in class.
Machine selection	Challenge supplier to meet best in class.	Long term, have supplier use best in class equipment.
Capital investment	Obtain and challenge equipment cost breakdown.	Purchase equipment from outside vendors.
	Show cost target equipment details.	Use LCC (low cost country) equipment vendors.
Tooling and its maintenance	Verify which tooling, if any, is amortized in piece price. Determine tool life. Obtain and challenge tooling cost breakdown.	Source tooling in LCC (low cost country). Eliminate charges for design on duplicate/replacement tooling.

(Continued)

TABLE 3.6 (*Continued*)
Negotiation Strategies and Potential Cost Reduction Opportunities

Type of Gap Driver	Negotiation Strategies	Potential Cost Reduction Opportunities
Cycle time	Obtain and challenge motion study breakdown for each station. Ask for calculation method. Verify in person during plant visit.	Shorten cycle time (i.e., reduce tool changes or times, increase machining speed, change materials if possible, etc.). Optimize utilization of mold space to fit more parts.
Manning	Obtain and challenge process layout and motion study for each operator. Verify in person during plant visit.	Reduce number of operators (i.e., combine stations, automate if justified, etc.). Eliminate shifts.
OEE/efficiency	Obtain root cause analysis of downtime. Challenge inefficiencies in the process (scheduled and unscheduled downtime, scrap).	Improve uptime (reduce scrap, reduce unplanned down time, use temps during lunches and breaks, etc.). Reduce scrapped process time.
Manufacturing overhead	Obtain detailed indirect allocation analysis.	Allocate indirect cost accurately (per activity-based costing methodology). Optimize electricity usage. Reduce number of support staff.
Capacity utilization	Challenge under-utilization. Pay only for utilization used by your product.	Improve equipment utilization. Increase number of shifts (i.e., 6-day pattern).
SG&A	Provide supplier with market data (i.e., financial statements).	Allocate SG&A costs accurately.
R&D	SG&A allocation should already include core R&D. Application specific R&D is under ED&T (engineering development & testing) category.	Eliminate any application specific engineering cost from R&D quote.
Profit	Provide supplier with market data (i.e., financial statements).	Lower to BIC (best in class) profit levels.
Cost of capital	Challenge interest rates. Only allow this cost on investment, not on expenses.	Apply actual interest rates instead of internal hurdle rates.
ED&T	Make sure piece price amortization assumes correct spend and life. Challenge overall spend.	Reduce overall ED&T cost (eliminate tests, reduce resources, etc.).
Payment terms	Challenge piece price impact of various payment days. Make sure interest rate is calculated properly.	Reduce payment terms (days). Reduce interest rates on working capital.

(*Continued*)

TABLE 3.6 (*Continued*)
Negotiation Strategies and Potential Cost Reduction Opportunities

Type of Gap Driver	Negotiation Strategies	Potential Cost Reduction Opportunities
Risk	Profit is already a reward for risk; no additional risk cost should be assumed.	Eliminate risk cost.
Startup cost	Separate and challenge startup costs.	Reduce startup cost.
Outbound logistics & packaging	Use logistics cost model to verify supplier quoted cost for competitiveness.	Optimize outbound logistics cost. In-source logistics to company provider.

To prepare for such a negotiation, below are some key preparation points. At the same time, the supplier's sales rep and/or cost estimator will be preparing for the negotiation, and many of the same key points are applicable.

- Gather all the necessary data such as cost breakdowns, manufacturing process details, value stream map, etc.
- Align on the documents that are to be given and/or shared
- Identify roles and responsibilities for each member of the negotiating team
- Role play the negotiation strategy to address the top cost gaps
- Anticipate supplier/customer responses and prepare answers for the possible scenarios
- Make sure the team is aligned on approach prior to negotiation
- Work with the buyer or the sales representative to identify what leverage might be there or can be created prior to negotiation so that the other side can be motivated to negotiate
- Understand supplier/customer motives and what they want to get out of the negotiation
- Establish atmosphere of collaboration and mutual respect during the negotiation
- Explain mutual benefits that could be achieved (e.g., expanding cost reduction findings to other products or customers)
- Lead the discussion, control the agenda, and stay on task
- Be prepared for various supplier/customer tactics, understand all the cost details
- Make it a data-driven negotiation, leave opinions out of it
- Let the buyer/sales rep handle purely commercial issues (e.g., profit, future business, etc.)
- Negotiation is built on compromise, both sides must feel like they won

SUPPLIER NEGOTIATION TACTICS

Customers should be watching for some common tactics (listed below) used by suppliers in order to confuse or to hide information in a non-collaborative relationship.

As a supplier, these are the tactics that should be employed to improve chances of success in a cost-based non-collaborative negotiation.

- Forcing the other side to defend their cost estimates
- Hiding or distorting manufacturing process information during plant visit (e.g., excess operators, increased cycle times, or rushed through tactics)
- Providing cost breakdowns that do not reflect actual cost
- Hiding excess profit in overhead, raw material, purchased components, or other cost categories
- Claiming ignorance in how costs are allocated
- Providing no information on accounting practices especially when it comes to overhead
- Showing only handpicked processes and avoiding others
- Not allowing to gather plant wide information such as factory size or total number of employees that would allow for reverse engineering of their P&L
- Not allowing access to their staff such as accountants, buyers, or anyone else that might know anything about their cost

SUPPLIER MANUFACTURING PLANT VISIT TIPS

One of the most important first steps for the customer cost estimator and buyer in any negotiation is the supplier manufacturing plant visit. The factory is where most of the costs actually occur. That is also why suppliers will often try to avoid having customers visit or try to keep the visits very short and focused. As a customer, the below tips are helpful to make such factory visits as productive as possible. As a supplier, these are some of the tactics to watch out for or to use in order to hide information from the customers.

- Ask the supplier to fill out a form with key manufacturing data and provide plant layout prior to the visit
- Prepare those visiting with you on expectations and their roles (e.g., counting operators, measuring cycle time, splitting up effort by processes or products, etc.)
- Start the supplier plant visit by presenting the objectives and provide a brief explanation of your function
- Optimize the plant touring time; don't let the supplier waste time with long presentations or trips out to lunch or dinner
- Make sure that the supplier has knowledgeable personnel guiding you through the plant tour
- Ask questions of people on the manufacturing floor including operators
- Take the time at each manufacturing process, don't get rushed through it
- Some suppliers don't allow stop watches, so learn how to count cycle time without it

- It is almost always against the supplier's company policy to allow picture or video taking, so please ask the supplier for permission if you would like to do so
- Fill out a plant visit input form yourself and agree with the supplier on the findings prior to leaving the supplier
- Record any potential cost reduction opportunities that might be possible; make the supplier aware of these if appropriate
- Give the supplier some general comments about the findings and thank them for their time and collaboration

MOCK NEGOTIATION TRAINING

It is not unusual for a company to have their cost estimators and buyers completely unprepared and untrained for cost-based negotiation. There even have been cases where companies have tried to implement collaborative cost-based negotiations with their suppliers, but had to drop the initiative due to lack of personnel that could actually perform the job. This is why it might be necessary for a company to develop its own cost-based negotiation training. This type of training should cover all the topics covered in this book, but it is often helpful to simulate the negotiation itself prior to the real one. The below mock negotiation exercise achieves this task and can be easily adjusted to reflect other situations. In this exercise, the trainees are split up into two groups, one representing a customer and the other representing a supplier. After being given a set of instructions (as below), they define their negotiation strategy and begin the negotiation process until a price is achieved satisfactory to both sides. Tables 3.7 and 3.8 can be used by each side to record pricing agreed upon at each stage.

TABLE 3.7
Customer Negotiation Tracker

Current quote	
Highest price that you are willing to pay	
Highest price that you would be happy with	
Max. price that you believe the supplier is willing to sell at	
Supplier's first counter offer	
Supplier's second counter offer	
Supplier's third counter offer	
Final agreed to price	
Any other cost reductions that were agreed to	

TABLE 3.8

Supplier Negotiation Tracker

Current quote	
Lowest price that you would sell at	
Lowest price that you would be happy with	
Max. price that you believe your customer is willing to pay	
Customer's first counter offer	
Customer's second counter offer	
Customer's third counter offer	
Final agreed to price	
Any other cost reductions that were agreed to	

INFORMATION FOR THE CUSTOMER SIDE

1. Your company, SPARK, has had trouble leveraging the connector suppliers, because most of them are large global corporations with established pricing. Only a handful of suppliers are able to manufacture this part due to its complexity and high tolerances.
2. Mold Experts Inc.'s first quote for this connector has come in lower than all of your current suppliers (5% lower than the next lowest supplier), but not low enough to justify it to the company to diversify further.
3. You need to negotiate with Mold Experts Inc. to reduce their price by another 30%. Otherwise you will not be able to sell them on rest of your organization, especially since they are regional and have no global presence. With 30% further reduction, the Mold Experts Inc. price would be low enough even if the parts had to be transported across the ocean.
4. Keep in mind that it's the full program spend that's up for negotiation, not just the first-year price.
5. Previous commercial negotiation failed to produce further price reductions from Mold Experts Inc. It's up to the cost engineer to evaluate the supplier's actual cost and find cost reduction opportunities if any exist.
6. You need this supplier to work out in order to meet your cost reduction targets. There are no other suppliers that can offer you savings. This is your last chance to close the deal. Otherwise there won't be enough time to execute the sourcing.

INFORMATION FOR THE SUPPLIER SIDE

1. Your company, Mold Experts Inc., is desperate for this job, because it is trying to grow globally within the automotive industry. Its previous efforts to get on the radar of large automotive companies have failed; however, it is doing quite a bit of business with smaller players.
2. Most of your current business is low volume with short program life, which drives your costs up. However, your costs for each product are not well

understood, since the controller usually assigns costs based on a rough calculation using plant wide rates developed using last year's budget numbers.

3. You would like to grow globally but your investors are not willing to put up the cash unless they get a three-year payback, so you quoted gross margins relatively high at 50% in order to have your customers fund your expansion. You are able to reduce your margins down to 30% if needed.

4. In the cost breakdown that your customer requested, you could not show your margins to be higher than 12%, so you hid them in other cost buckets such as raw material and overheads.

5. Since SPARK also requires suppliers to provide productivity givebacks, even though any process improvements are usually eaten away by inflation on labor/energy/materials, you had to build the givebacks into the first-year price. However, the customer does not allow you to show this in the margin, so you also had to hide productivity givebacks in other cost buckets.

6. Based on your research, the price you offered is competitive. It's not clear how the SPARK cost engineer will approach the negotiation since you've never met with him/her before. However, the plan is not to give any consolations unless the cost engineer has some real data to prove your assumptions wrong.

7. This is your last chance to close this deal. There will be no further rounds of negotiation.

SUMMARY

The cost-based negotiation is very different from the typical transactional or high-stakes negotiations that are taught in Karrass and Harvard Law School trainings, respectively. Although the concepts of both Karrass and Harvard are also applicable, the cost-based negotiation is data driven and focuses on finding drivers of cost differences between customer and supplier assumptions, then collaborating in problem solving to eliminate those differences either through commercial negotiation or through cost reduction activities.

There are many intricacies of cost-based negotiation that require high levels of expertise from the participants. For example, instead of simply choosing the lowest quote, the customer's buyer and cost estimator will need knowledge of the supplier's manufacturing processes and cost structures. This knowledge is best acquired through supplier manufacturing plant visits where customer representatives can visualize the size of the factory, number of operators, number of support staff, flow of material through various manufacturing processes, and many other cost drivers. These can be used to develop cost models simulating the supplier's P&L and pricing.

Considering that there is no publicly available training for cost-based negotiation, companies must develop their own resources, otherwise the company's staff will struggle negotiating based on cost data. This book offers a lot of help on developing such training, the most useful of which is the mock negotiation exercise which simulates cost-based negotiation prior to the real one.

4 Cost Estimating for Various Manufacturing Processes

Cost estimating is an extremely important capability that every company needs; however, it is difficult to master due to its various intricacies. This is why it is more rare than common for a company to have this capability. No studies have ever been published on this topic, but it would be of no surprise if a large amount of bankruptcies were due to the lack of accurate cost estimating. Thus, a company with this capability would have a huge competitive advantage and should strive to maintain and grow this capability. It is an absolute must for an efficient cost engineering process. Without it, companies will struggle to understand cost and to make any improvements to it. A company must first know its cost before it can do anything about it. This is why an additional chapter is dedicated to estimating basics for various manufacturing processes.

Every manufacturing process requires a lot of experience to fully comprehend and master, but it is often important for a manufacturing professional – whether it is someone in engineering, purchasing, operations, or other company functions – to have at least a rudimentary understanding of all basic manufacturing processes and how they relate to cost. For any individual who is part of the company's cost engineering process, there will most likely be a need to understand the cost of more than just one manufacturing process. This is especially true for companies that only assemble purchased components that are produced by suppliers employing various other manufacturing processes. The customer's buyer or cost estimator must be able to understand cost of manufacturing processes that are not that customer's core competency, so he/she can estimate their purchased cost and negotiate on that basis.

For example, an electronic control unit (ECU) assembly will have electronic components such as micros or resistors assembled to a PCB (printed circuit board) using SMT (surface mount technology). It might also be composed of a molded plastic housing that will probably have copper pins/terminals that are stamped then plated, and it even might have an aluminum cover and heat sink that are die casted or stamped. These are at least five basic manufacturing processes involved in making just one assembled component.

This chapter will aim to provide some basic cost estimating considerations for the manufacturing processes such as sand casting, die casting, stamping, forging, powder metal forming (sintering), machining, molding, painting, coating, plating, SMT, custom assembly, electronic component manufacturing, and additive manufacturing.

The basic concepts that follow are not meant to be comprehensive insights of each manufacturing process, but rather to give an overview of the main cost drivers. Since manufacturing processes are constantly evolving and raw material pricing is constantly fluctuating, the estimates shared are a rough guide only and do not necessarily represent every manufacturing scenario. There will be many outliers that will not fit within the general range of guidance numbers. The main takeaway should be how cost is developed for each of the manufacturing processes. Cost estimators should develop their own estimates and cost models based on the unique set of variables applicable to their parts that represent economic realities of their specific manufacturing processes employed.

A good reference book for additional detail is "Realistic Cost Estimating for Manufacturing," edited by Michael Lembersky and published by the Society of Manufacturing Engineers (SME).

SAND CASTING

The primary cost component for this manufacturing process is the raw material, which is mainly either grey or ductile iron with the difference being their chemical composition. The differentiator is the amount of graphite used with ductile iron having more than the grey iron. Each iron has its own properties that are suited for different applications. The main components are basically the same with pig iron (pure iron from the mine formed into billets) and metal scrap, both of which are sold on the metal exchange markets. The mix between the two components can vary, but a 50/50 split is a safe assumption to start with. For example, pig iron per ton might cost $300 and scrap metal per ton might cost $400, so if the two were melted together, the cost per ton would be about $350.

It is important to note that there will be some material losses that will occur during melting due to oxidation. In layman's terms, oxidation is like the froth that collects on top of a chicken soup when it is cooked, and it needs to be skimmed off and thrown out. The same thing happens when melting iron, and this results in material losses of 2%–7% (average of 5% is a good starting assumption) depending on batch chemistry.

Upon arrival at the casting supplier, both types of material are stored in large pits from which they are later collected by a large overhead magnetic crane operated by a single person and placed into massive induction melting furnaces (30–60-ton capacity). These furnaces consume large amounts of energy – either gas or electricity – and are never turned off except for periodic maintenance. Although they must operate 24/7 to keep molten iron from solidifying and seizing up the equipment, they are not usually actively producing products 24/7. This is a normal characteristic of the process, not an inefficiency of the supplier, so an allowance must be made for this natural production down-time when determining furnace's hourly cost rate that can reach $100–$300 per hour depending on the furnace size, which includes the cost of energy, equipment, space, two or three operators, and various other things. The hourly cost rate for the furnace should be determined by dividing the total furnace operational cost over total production hours (perhaps 20 hours/day, 6 days/week, and 50 weeks/year or 6,000 hours per year). For example, if the cost of operating the furnace is $600,000 per year, then the hourly cost rate would be $100 ($600,000/6,000 hours).

The next step in the process is to pour the molten iron into giant ladles transported by an overhead crane. These ladles of molten iron are destined for a specific product; thus, their chemistry needs to be adjusted to make sure the material properties are correct. This usually means adding whatever chemical components that might be lacking such as carbon, silicon, or other metals. The original scrap that is melted together with pig iron will have a random chemical composition, so its chemistry will be varied with each batch. These added chemical components are usually very small in quantity, so although they are more expensive than iron, they do not significantly affect the overall raw material cost.

After the required chemical composition is established, the ladle crane operator will move batches of molten iron to the area where it will be poured into a pattern. The pattern is a part shape imprinted into wet sand mixed with adhesives and held together inside a box. This part of the process is different for low- and high-volume products. The low-volume process would involve manually producing a pattern in several steps then manually pouring the molten iron into the mold cavity created by the pattern. For this book, we will focus on the high-volume production process, which is automated. A typical automated casting process would be something like the one made by the DISAMATIC® (registered trademark of DISA Industries A/S), as shown in Figure 4.1, a well-known manufacturer of casting equipment.

It starts with creating a mold by the automated DISAMATIC® machine that makes sand molds using two pattern plates imprinting in sand both sides of the desired part shape, as shown in Figure 4.2.

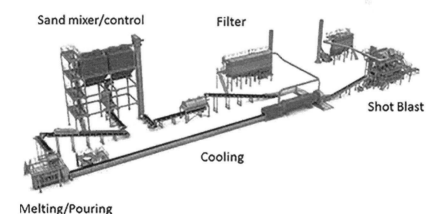

FIGURE 4.1   DISAMATIC® casting process. (Courtesy of DISA Industries A/S.)

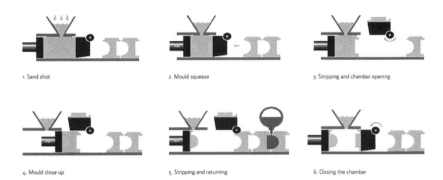

1. Sand shot   2. Mould squeeze   3. Stripping and chamber opening

4. Mould close up   5. Stripping and returning   6. Closing the chamber

FIGURE 4.2 DISAMATIC® mold creation process. (Courtesy of DISA Industries A/S.)

FIGURE 4.3 DISAMATIC® mold pattern plate. (Courtesy of DISA Industries A/S.)

The two halves are imprinted separately, then put together to form a cavity inside of it, which is exactly in a shape of the casted part or multiple parts depending on design. Figure 4.3 is a picture of one of the DISAMATIC® mold pattern plates for a car brake rotor.

Since a part like the rotor might have some design features inside of it, there could be a need for a core to be inserted into the cavity that has those internal features. The core would be made out of the same wet sand but will probably need to be coated (thus the white color of the cores in the example in Figure 4.4) to maintain its shape during the iron pouring process.

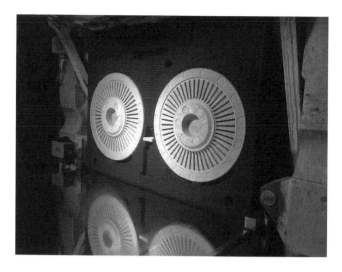

FIGURE 4.4 DISAMATIC® pattern with cores. (Courtesy of DISA Industries A/S.)

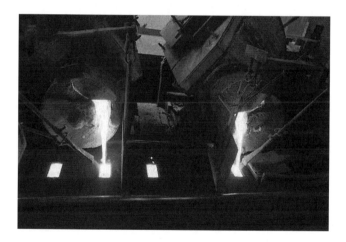

FIGURE 4.5 DISAMATIC® iron pouring process. (Courtesy of DISA Industries A/S.)

After the sand boxes are formed and cores inserted, the molten iron is poured into the cavities, as shown in Figure 4.5. The mold making and pouring are synchronized as part of one automated casting system. Only one or two operators would be needed at this operation to monitor the iron pouring. A separate operator will be needed to manually insert the cores into the mold cavities, but this process could be automated using a robot.

FIGURE 4.6   Four-cavity mold. (Courtesy of DISA Industries A/S.)

The DISAMATIC® machine then pushes the sand molds through a long cooling line where the molten metal cools off inside the mold cavities to form a shape. The cooled off mold (e.g., four-cavity brake rotor mold shown in Figure 4.6) is dropped into a tumbling shaker to separate it from the sand (which is sent back on a conveyor line to be reused).

The sprue, runners, and risers, which are part of the mold, are also separated from the parts and sent back for re-melting as part of the scrap. Depending on the part design, the de-gating will require one or more manual operators. After this, the parts are usually taken to separate manually operated deburring and/or sandblasting stations after which the parts are ready to be shipped to the customer. The customer might also require some testing prior to shipping to make sure there are no cracks or porosity in the casted parts.

The above described casting process is very capital intensive with each system costing $5–$10 million depending on size; therefore, it is important to optimize the number of parts produced per hour or per day using the system. The throughput is very dependent on part design because that drives the number of cavities that can fit in a single mold and it also drives the pouring time per cavity. The key to optimizing casting cost then is optimizing the design to achieve the highest number of cavities per mold and the shortest pouring cycle time (which is a bottleneck driving the cycle time of the whole system).

To summarize, Table 4.1 describes the key cost drivers for a sand casting process. The provided cost estimates could vary widely depending on multiple factors and can exceed given ranges.

TABLE 4.1
Sand Casting Cost Drivers

Key Cost Driver	Description	Cost Basics
Raw Material	Pig iron mixed with steel scrap Sand	~$0.15–$0.25/lb. including delivery to the foundry assuming 50/50 split between pig iron and scrap plus cost of other special chemicals ~$0.01/lb. of iron poured, most of the sand is recycled but there is cost of recycling and losses due to overbaking
Scrap	Loss of material only about 5% due to oxidation; everything else is re-melted Process scrap is usually 2%–5%	~5% of raw material cost ~2%–5% of process cost (material is recycled)
Core Making	Molding press with 1 operator or robot molding wet sand Secondary operation coating the core with adhesive	~$0.40 (brake caliper) – $1.50 (brake rotor) per core depending on shape and size
Melting	Large gas or electric fired furnace (~$0.5–2.0 M) with 2 operators with several tons of iron melted per hour Ladle system to pick up molten iron and deliver it to casting machine	~$0.05–$0.10/lb
Casting	DISA type casting machine/system (~$5–$10 M) with 4–10 operators depending on part design Cycle time per mold ranges between 40 and 60 sec depending on mold size and part design	~$2.50–$6.00 per mold depending on size of casting system Cost per part depends on how many parts are casted per mold
Tooling	Part-specific patterns for molding and core making	~$100–$300k depending on pattern size and part complexity

REAL-LIFE ANECDOTE

A casting buyer was not able to comprehend the trends in supplier pricing for castings, so he imposed on his suppliers a cost per pound rule where, regardless of design, the price was determined by multiplying weight by a cost per pound target from the buyer. This target was usually some average of existing pricing, therefore the suppliers found themselves in a situation where only the smaller castings were meeting the target requirement. This was because more castings could be molded per pattern for smaller parts thus reducing the processing cost per piece. The large parts usually were able to accommodate only two or even one cavity per mold, thus increasing processing cost per part. The result of this buyer's policy was that the pricing of his smaller parts increased over time closer to his average, while he couldn't find anyone to quote the larger parts. Fortunately, the company's cost estimator for castings was able to develop a simple pricing model for the buyer where material and processing costs were separated and used different cost drivers to determine overall pricing. The pricing model reflected the suppliers' costing realities where material piece cost is based on weight and processing piece cost is based on design and ultimately cavitation and pouring time.

DIE CASTING

This process is similar in many ways to plastic molding because of its use of pressure to "squeeze" heated raw material (such as molten aluminum or magnesium) into shape (see diagram of the process in Figure 4.7). It is also similar to sand casting in that the process involves a melting furnace and post casting finish operations such as sand blasting, deburring, or trimming. The melting furnace size is much smaller than for the iron casting process, but still has a capacity of 20–40 tons and also must run 24/7 to maintain temperatures at a molten level.

The raw material, such as aluminum alloy, is delivered to the melting furnace in the shape of ingots, which are easy to store in a warehouse environment and

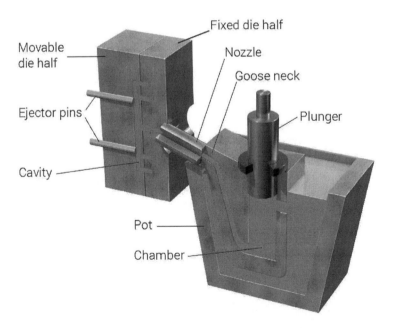

FIGURE 4.7 Die casting process. (Courtesy of engineeringclicks.com; https://www.engineeringclicks.com/die-casting/.)

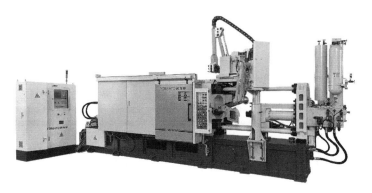

FIGURE 4.8   Die casting press. (Courtesy of Jiangsu Yomato Machinery Technology Co., Ltd.)

do not require overhead cranes. Usually, the melting furnace area of the factory will have two or three furnaces which are manned by one or two operators. The molten Al is then poured into ladles that are delivered by a Hilo to a smaller holding furnace located at the die casting press.

The most critical part of the die casting process is the actual die casting itself with press sizes ranging between 135 and 3,500 tons depending on part size. The press (see Figure 4.8) is usually operated by one person who is responsible for maintaining holding furnace load and temperature, proper operation of the die casting press, and any light post operations, while a robot usually performs the function of removing the casted parts out of the press. Sometimes it is also possible that the human operator is shared between two or more presses, but this is equipment and part design dependent. The same dependency is true for the cycle times, but usually the drop-to-drop time (time between parts being removed from the machine after completion of casting) is anywhere between 35 and 65 seconds. Keep in mind that there could be multiple parts being casted at a time, so the cycle time per part would be the cycle time per mold divided by the number of parts casted per mold (similar to the sand casting process).

The gates and risers or the channels through which molten Al is delivered to the part cavities of the tool have to be separated from the parts and removed. Fortunately, these can be re-melted and the material reused for casting, so there is very little material loss. The only material loss occurs due to oxidation, sometimes called dross loss. Similar to the sand casting process, the dross is like the "soup froth" that collects on top of the molten Al that has to be skimmed off and thrown away.

After the casted parts are given time to cool off, they might need secondary operations such as deburring of rough edges, sand blasting to clean surfaces, or trimming to cut off any larger burrs.

To summarize, Table 4.2 describes the key cost drivers for a die casting process. The provided cost estimates could vary widely depending on multiple factors and can exceed given ranges.

TABLE 4.2
Die Casting Cost Drivers

Key Cost Driver	Description	Cost Basics
Raw Material	Pre-melted alloyed ingots mixed with internal scrap	~$0.50–$1.50/lb. for Al ~$3.00–$4.50/lb. for Mg
Scrap	Loss of material only about 3%–5% due to oxidation; everything else is re-melted Process scrap is usually 2%–3%	~3%–5% of raw material cost ~2%–5% of process cost (material is recycled)
Melting	Gas or electric fired furnace (~$0.5–$1 M) with 1 operator per 2–3 furnaces Molten Al poured into a ladle that is moved to the press by a Hilo	~$0.05/lb.
Casting	Casted in 135–3,500-ton presses ($350–$2,500k) with an operator manning usually one and sometimes two presses Cycle times depend on press sizes and part designs, but range between 35 and 65 sec per mold	~$1.50–$4.50 per mold depending on size of casting system Cost per part depends on how many parts are casted per mold
Tooling	Part-specific dies	~$100–$300k depending on pattern size and part complexity

FORGING

The forging process involves using presses to hammer a billet into a desired shape, defined by the shape of tool cavities, with multiple hits (see closed die forging in Figure 4.9). Depending on part size and shape, it can be cold (smaller parts) or hot forged (larger parts). The only difference between the two is that in the hot forge process the billet is heated up to be more malleable.

Press sizes and their cost can vary greatly depending on part size and forging process used. Figure 4.10 shows basic design of a Sumitomo Heavy Industries forging press.

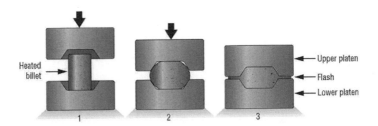

FIGURE 4.9 Forging process. (Courtesy of forgemag.com; https://www.forgemag.com/articles/84766-micro-cold-forging-helps-extend-die-life.)

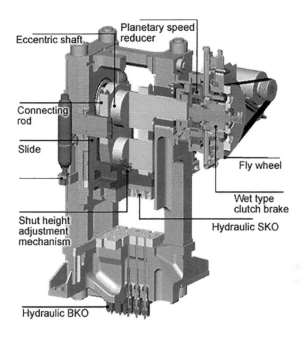

Planetary speed reducer

Eccentric shaft

Connecting rod

Slide

Fly wheel

Wet type clutch brake

Hydraulic SKO

Shut height adjustment mechanism

Hydraulic BKO

FIGURE 4.10   Forging press. (Courtesy of Sumitomo Heavy Industries, Ltd.)

TABLE 4.3
Forging Cost Drivers

Key Cost Driver	Description	Cost Basics
Raw Material	Pre-melted alloyed ingots formed into billets	~$0.30–$1.00/lb. for steel
Scrap	Loss of material due to flash; surface oxidation and other factors could be as high as 30%	~20%–30% of raw material cost
	Process scrap is usually 3%–5%	~3%–5% of raw material and process cost
Forging	Forged in 100–16,500-ton presses ($350–$4,500k) with an operator manning usually one press Cycle times depend on press sizes and part designs, but range between 3 and 25 sec per part	~$50–$900 per hour depending on press size
Tooling	Part-specific dies	~$100–$300k depending on pattern size and part complexity

To summarize, Table 4.3 describes the key cost drivers for a forging process. It is also important to account for all the secondary operations associated with forging such as inventory movement and storage. The provided cost estimates could vary widely depending on multiple factors and can exceed given ranges.

MACHINING

This is probably the most complex of all manufacturing processes to cost estimate. There are thousands of machine choices, and the choice is often dependent on what machine is currently available on the production floor at any given supplier and not necessarily on what is the most optimal machine to use. The part design could also be very complex with dimensional tolerances and surface specifications potentially driving very complex machining steps with cycle times often difficult to determine. Despite these complexities, the most common machining center is the CNC (computer numerical control) similar to the one shown in Figure 4.11, which gives companies the most manufacturing flexibility.

The key drivers of the machine selection and the resulting cycle times are "speed and feed," meaning machining tool rotational speed and the feed rate at which the tool is pushed through the material in order to remove it from a part (see Figure 4.12). There are many books written on this subject, so the "speed and feed" calculation method will not be covered in this book, but it is determined by types of material

FIGURE 4.11   CNC machine. (Courtesy of Okuma America Corporation.)

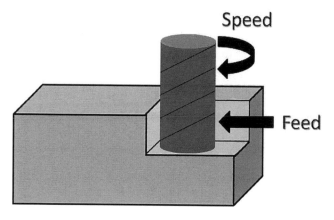

FIGURE 4.12 Machining speed and feed diagram.

TABLE 4.4
Machining Cost Drivers

Key Cost Driver	Description	Cost Basics
Raw Material	Almost all types of material can be machined including iron, steel, Al, and plastic There is no other material used during the machining process outside of the material being machined	Depends on material being machined Must consider gross weight of material less what is removed (sold as scrap)
Scrap	Material that is removed from a part is usually recycled and resold on the scrap market During machining, some parts are scrapped due to tool wear or breakage	~2%–3% of raw material and process cost
Machining (e.g., turning, grinding, etc.)	Various machines can be used including CNCs, transfer lines with multiple CNCs, machining centers Each machine can have one or more spindles, meaning that one or multiple parts can be machined at the same time	Depends on machine type and cycle time developed based on "speeds and feeds," which in turn depends on material machined and desired surface finishes
Tooling	Perishable tools that have diamond or other tips which wear with use	Highly dependent on number of tools used to achieve part finishes, but usually cost of perishable tools is amortized in piece price

being machined, its hardness, the machine and tool used, and the cooling system in place (e.g., oil or other specialty fluids).

To summarize, Table 4.4 describes the key cost drivers for a machining process. The provided cost estimates could vary widely depending on multiple factors and can exceed given ranges.

POWDER METAL SINTERING

Although the term "sintering" suggests that the powder metal (PM) is only some-how heated to shape, the PM processing actually consists of several processes (see Figure 4.13). First, the PM is mixed and molded into shape on a large press, then the parts are sintered, meaning they are put through a heating furnace, where the PM granules "melt together," then the parts are sized, or trimmed, to assure dimensional requirements that might have been loosened during the sintering process. The parts might also be tested for dimensional and functional requirements.

To summarize, Table 4.5 describes the key cost drivers for a powder metal sinter-ing process. The provided cost estimates could vary widely depending on multiple factors and can exceed given ranges.

THE POWDER METAL
COMPACTION PROCESS

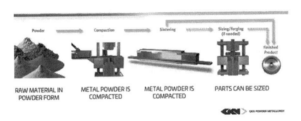

FIGURE 4.13   Powder metal compaction and sintering process. (Courtesy of GNK Powder Metallurgy.)

TABLE 4.5
Powder Metal Sintering Cost Drivers

Key Cost Driver	Description	Cost Basics
Raw Material	Metal in the form of granules of 1–3 mm in diameter	~$0.60–$1.50/lb.
Scrap	Since parts are pressed into shape, there is no material scrapped for each part (no offal)	~2%–3% of raw material and process cost
	However, there are parts scrapped during the processes usually due to tool wear or incorrect sintering temperatures	
Pressing	40–200-ton presses (~$500–$1,500k) are used that force material into shape defined by a tool (usually one cavity tool)	~$0.10–$0.50 per part depending on part and press size
	One operator can usually run one or two presses with part cycle times between 5 and 7 sec	
Sintering	This process consists of simply putting the pressed parts through a large heating furnace (~$1 M)	~$0.05 per part
	One or two operators load and unload the parts with cycle times around 5 seconds per part	
Sizing	Small press (~$100–$300k) operated by one person	~$0.05 per part
Tooling	One cavity tool	~$20–$60k depending on part design

STAMPING

One of the most widespread manufacturing processes, stamping is used to make parts for a multitude of purposes. It can range from very small parts, such as door hooks, all the way to things like car panels. Similarly, the machines used to make these parts can vary dramatically in size, but the stamping process can boil down to two main categories: transfer and progressive die (see progressive die in Figure 4.14).

The transfer die process, usually used for larger parts, consists of large sheets of metal being loaded and unloaded in multiple presses either manually or by a robot in order to achieve final shape. On the other hand, the progressive die process is usually used for smaller parts and involves a coiled strip of metal fed automatically into the press with multiple tool stations, each station stamping out a different feature until the final part shape is achieved (see Figure 4.15).

The parts might also require different post stamping operations depending on part design and requirements. De-burring, sand blasting, polishing, etc. are the most common. A part might also require a secondary bending operation to get it into the final shape. This is usually done on smaller offline presses. Also, stamped parts are often welded together into assemblies. This is typically done in booth cells with human or robot operator.

FIGURE 4.14 Progressive die press. (Courtesy of Pacific West America, Inc.; https://pacificwestamerica.com/product/custom-metal-stamping/250-ton-stamping-press-1/.)

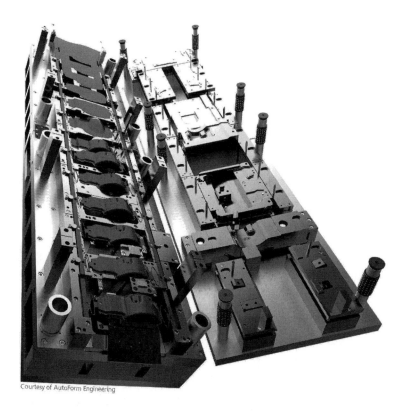

Courtesy of AutoForm Engineering

FIGURE 4.15   Progressive die tool. (Courtesy of AutoForm Engineering; https://www.autoform.com.)

To summarize, Table 4.6 describes the key cost drivers for a stamping process. The provided cost estimates could vary widely depending on multiple factors and can exceed given ranges.

PLATING (ELECTROPLATING)

Most of the time, the casted, forged, machined, sintered, or stamped parts require some kind of environmental protection because the parts are often used in an open environment and the processed metals do not have inherent environmental protection. The metal surfaces are exposed during delivery and usage, so they have to be coated or plated in order to prevent rust and deterioration over time.

There are three main ways to plate parts: barrel plating, rack plating, and coil plating. The first, barrel plating, usually meant for smaller parts, involves rotating a cage full of parts inside multiple chemical baths (see Figure 4.16). Because parts are

TABLE 4.6
Stamping Cost Drivers

Key Cost Driver	Description	Cost Basics
Raw material	The material comes pre-formed from the metal suppliers; it can come in sheets or coils and can vary dramatically in sizes There are also various thicknesses, but to be able to form the metal without heating it up limits the thickness to about 5 mm	~$0.25–$0.50/lb. for steel ~$0.50–$1.50/lb. for Al ~$2.50–$4.50/lb. for copper and copper alloys
Scrap	The percentage of parts scrapped during the process is usually fairly low, but the amount of material that is scrapped for each part as it is stamped into the final form (offal) can be very large (up to 70%) depending on part shape Proper nesting of parts in the working piece is key to minimizing offal However, all scrapped material can be resold for re-melting (at about 25%–80% of its original value depending on material)	~2%–3% of raw material and process cost ~10%–70% of used material (offal or difference between gross and net weight)
Stamping	Wide variety of presses with sizes between 20 and 1,500 tons (~$40k–$2 M) Each press is operated by one or two operators or by robots depending on the part and press used Cycle times are usually very low, usually 2 sec or less, and tools can be designed to stamp out multiple parts at a time	~$0.01–$0.30 per part depending on part design and press size, but overall the process cost per part is very low compared to the material cost
Tooling	Many variations of tools with single or multiple cavities	~$40–$200k depending on part design

thrown together into a cage and hit each other as the cage rotates, the parts should not have any critical features that could be damaged during this process.

The rack plating process, meant for larger parts, also requires parts to be dipped in chemical baths, but due to shape, weight, or size, the parts are first hung on a rack (see Figure 4.17), which is then dipped in multiple baths. The thickness of the plating material is controlled by the level of electrical current passed through the parts. This way, only a certain amount of particles (e.g., zinc) with opposite charge attach to the part. Masking is sometimes utilized to cover surfaces that require un-plated surfaces to remain.

Finally, the coil plating (see Figure 4.18), sometimes called reel-to-reel plating, is meant for parts still attached to a coil (e.g., partially stamped out connector pins in Figure 4.19). This involves running the coiled parts through several smaller baths.

FIGURE 4.16 Barrel plating. (Courtesy of H&E Plating, Ltd.; https://heplating.com/plating_methodsbarrel_electroplating.)

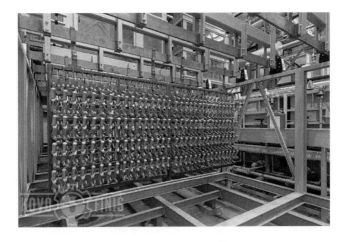

FIGURE 4.17 Rack plating. (Courtesy of KOVOFINIŠ s.r.o.; https://www.kovofinis.cz/en/electroplating.)

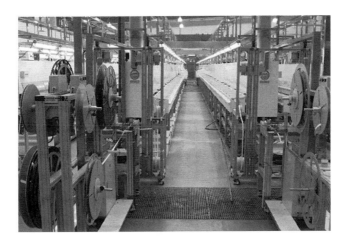

FIGURE 4.18 Coil plating. (Courtesy of Interplex; https://interplex.com/resources/plating/technical-bulletins/reel-to-reel-plating-improves-production-efficiency-and-saves-costs/.)

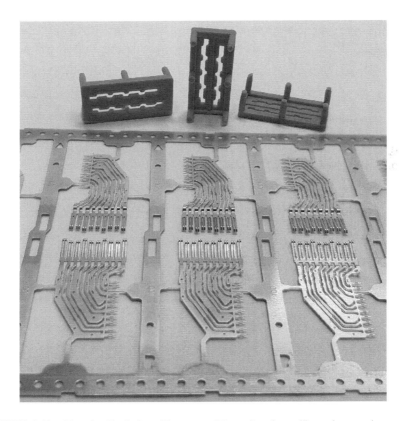

FIGURE 4.19 Plated coil of pins. (Courtesy of Interplex; https://interplex.com/resources/plating/technical-bulletins/reel-to-reel-plating-improves-production-efficiency-and-saves-costs/.)

TABLE 4.7
Plating Cost Drivers

Key Cost Driver	Description	Cost Basics
Raw Material	There are various types of plating materials, most with metallic bases such as zinc, nickel, silver, or gold	~$0.75–$1.50/lb. for zinc ~$5.00–$7.50/lb. for nickel ~$1.00–$1.50/lb. for chrome ~$15–$20/ounce for silver ~$1,000–$2,000/ounce for gold Cost per part depends on type of plating material, surface area and thickness
Scrap	There is no loss of plating material due to accuracy of electroplating process However, there is some scrap due to process issues that might occur and due to changeovers	~2%–3% of raw material and process cost
Plating	All plating processes require several baths including some for material cleaning prior to plating Typical equipment cost ranges between $0.5 to $3 M and is operated by 1 or 4 people (highest on rack plating where laborers must manually hang and un-hang every part on a rack hook) However, a plating facility will also require ~$1 M investment in water filtration and waste collection systems Cycle times range from 0.5 sec (coil plating) to 15 sec (rack plating) per part depending on equipment and part design Rack plating is also highly dependent on rack size which drives amount of parts possible per rack	~$0.005–$0.50 depending on process and part design
Tooling	Usually there is very little tooling involved, mostly some racks for rack plating	~$10–$20k for racks

The plating thickness is also controlled by the level of current passed through the coiled parts. The additional benefit is the ability to control vertically which portion of the partially stamped out coil is plated, meaning it can be the upper, lower, or middle section of the coil.

To summarize, Table 4.7 describes the key cost drivers for a plating process. The provided cost estimates could vary widely depending on multiple factors and can exceed given ranges.

MOLDING

The process of molding plastic is probably the most common manufacturing process in the world with different variations such as injection molding, blow molding, compression molding, and insert molding. Almost anything can be made from plastic resin. This is because of its flexibility of application as well as affordability of production. The molding manufacturing process really consists of only one step, which is pressing resin under pressure and heat into a shape defined by two halves of a metal tool (see Figure 4.20). The parts are then given some time to cool off after which the tool opens and the finished parts drop down to a bucket below or are removed by a robot arm.

In the case of insert-molding or over-molding, robot arms will be used to insert metal components into the tool before molding, then remove finished parts from the tool after molding, before finally placing them into packaging or into testers (e.g., electrical test for over-molded copper pins).

To summarize, Table 4.8 describes the key cost drivers for a molding process. The provided cost estimates could vary widely depending on multiple factors and can exceed given ranges.

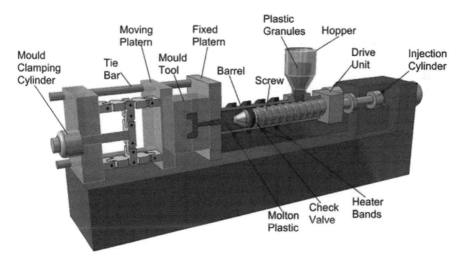

FIGURE 4.20 Injection molding process. (Courtesy of Rutland Plastics, Ltd.; https://www. rutlandplastics.co.uk/plastics-moulding-methods/moulding-machine/.)

TABLE 4.8
Molding Cost Drivers

Key Cost Driver	Description	Cost Basics
Raw material	There are thousands of various resins and their grades, tuned in to different requirements for flexibility, hardness, endurance, durability, and temperature. These are met by various additives that alter the properties of basic resins The most well known additive is glass fiber which improves durability and temperature resistance (e.g., 30% glass filled [GF] has a content of 30% glass fibers) All resins come to the molder in a granule form in bags, on pallets, or uploaded to large holding silos	~$1.50–$2.50/lb depending on type of resin and amount of quantities purchased Prices can go up dramatically if only small quantities are ordered annually or per shipment
Scrap	The parts themselves can usually be molded to finished (net) shape; however, the resin that solidifies in the delivery channels of the tool must be scrapped Depending on part and tool design, this resin can be an additional 10%–60% of part finished weight Fortunately, it is often allowable to reuse this scrapped material after grinding it into granules (re-grind); otherwise, it can be re-sold at about 10% of its original value The scrapping of bad parts is usually very low, perhaps only 0.5%–2% depending on part design and requirements	~0.5%–2% of raw material and process cost ~10%–60% of shot weight material (gates and risers) unless it can be re-ground into granules and reused for molding
Molding	Various press sizes (40–600 tons) are used depending on application The cost of these presses depends on size, make, and auxiliary equipment needed, but usually ranges between $100k and $1 M For simple molding operations where parts drop into a bucket, 1 operator can run up to 4 or 6 presses For insert or over-molding processes, an operator or robot will be needed at each press Cycle time varies widely depending on press size, resin used, and part/tool design, but ranges between 15 and 65 seconds per mold Cycle time per part depends on how many parts can be molded per mold	~$0.05–$1.00 depending on press size and part design
Tooling	Highly dependent on part and tool design	~$40–$200k per tool

PAINTING

The painting process involves applying paint to plastic or metal parts. This is done either through spray nozzles operated manually by a person or by automated machines often employing robots (see paint booth in Figure 4.21). The paint application is usually completed in a booth with some kind of water system to collect the paint overspray.

For highly decorative parts, such as plastic bezels and buttons for a car radio/ HVAC controls a part will usually require several paint applications and thus multiple booths and drying ovens. The part might also require pad printing (rubber stamping) and laser etching in order to achieve a desired look and feel. Although painting is often considered an afterthought in manufacturing, this can be a very difficult process and usually requires specialized experts to avoid high scrap rates.

To summarize, Table 4.9 describes the key cost drivers for a painting process. The provided cost estimates could vary widely depending on multiple factors and can exceed given ranges.

FIGURE 4.21 Automated spray paint booth. (Courtesy of WOLF Anlagen-Technik GmbH & Co. KG; https://www.wolf-geisenfeld.de/en/industrial-spray-booths.)

TABLE 4.9
Painting Cost Drivers

Key Cost Driver	Description	Cost Basics
Raw material	There are thousands of colors available and many more can be generated by pre-mixing Also, there are many base and top coats available that might need to be applied prior to and/or after the main paint coat	~$40–60/gallon for base coat ~$60–150/gallon for paint ~$70/gallon for top coat Cost per part determined by surface area, thickness, and cost per gallon
Scrap	This can be a very significant cost driver in painting due to many factors Visual requirements from part to part can drive many paint applications and laser etching, each capable of generating rejects	~5%–30% of material and process This is calculated as a cumulative total, meaning that if there are 5 steps to paint the part, at step 2 the scrap would be material and process that is related only to steps 1 and 2

(Continued)

TABLE 4.9 (*Continued*)
Painting Cost Drivers

Key Cost Driver	Description	Cost Basics
	Also, the paint process itself is fairly inaccurate and a lot of overspray can be generated, which has to be collected (usually with water) and sent to specialized landfills	Also, the overspray for each painting step needs to be considered, which could result in 10%–30% of additional material
Painting	Depending on part and visual requirements, the process can be a manual single booth operation (~$500 k) with one person spraying with a single nozzle or a fully enclosed automated paint line (~$1.5–3 M) that has multiple paint booths with robot operated nozzles and multiple curing ovens	~$30–$200 rate per hour or $0.02–$2.00 per part
		~$0.02 per pad print
		~$0.02–$0.10 per laser etch depending on size
	These systems also require a water collection system for the paint overspray that is later compressed into solid blocks and sent to a landfill	
	Although the paint lines are automated, operators are still needed to hang and un-hang parts from fixtures. Depending on part design and line speed, 2–8 operators will be required	
	The cycle times vary widely depending on part design and systems used, but can be anywhere between 15 and 60 seconds per part	
Tooling	Usually only minimal number of tools are required, and these consist of fixtures on which parts will sit during painting	~$20–$100k for all fixtures depending on process

SURFACE MOUNT TECHNOLOGY (SMT)

The process of placing electronic components – such as micros, resistors, and capacitors – on a printed circuit board (PCB) (see Figure 4.22) is called the surface mount technology (SMT). In the age of electrification, SMT is one of the most common processes in manufacturing.

SMT is an automated process consisting of several in-line steps (see SMT line in Figure 4.23). First, the PCB panels have solder paste stenciled on them, then the electronic components are placed on the boards using one or multiple robot arms (sometimes called "pick and place") machines, then the boards are placed in the reflow oven where the solder paste is heated up so that it can flow around the connection pins of each electronic component, and finally the parts are inspected by cameras to assure proper component placement.

FIGURE 4.22 Printed circuit board assembly (PCBA). (Courtesy of Arrow Technical Services, Ltd.; http://www.arrowtechnical.com/services/volume-manufacture/surface-mount-pcb-assembly/.)

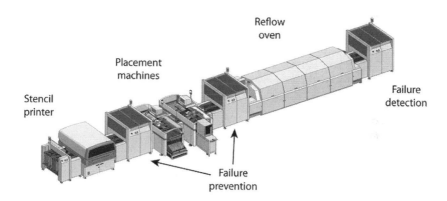

FIGURE 4.23 SMT process. (Courtesy of surfacemountprocess.com; http://www.surface-mountprocess.com/.)

There are also post-SMT processes required such as de-paneling and in-circuit testing (ICT). De-paneling is simply cutting the finished PCBs out of the panel using a simple press operated manually by one person. ICT is needed to make sure that all the circuits in a finished PCBA are operational, which confirms proper electronic component placement and correct soldering. This test machine (~$50–$150k) is operated by one manual operator, but the cycle time is dependent on the number of circuits and/or components that need to be tested.

To summarize, Table 4.10 describes the key cost drivers for an SMT process. The provided cost estimates could vary widely depending on multiple factors and can exceed given ranges.

TABLE 4.10
SMT Cost Drivers

Key Cost Driver	Description	Cost Basics
Raw Material	SMT is an assembly process and most of the material used is purchased components such as micros, resistors, and capacitors The only "raw" material is the solder paste and its cost is relatively small compared to the cost of purchased components	~$0.01 per square inch for solder paste
	One important consideration is the utilization of the PCB panel The shape of a PCB can vary widely, but these are all later cut out of a standard panel size The more PCBs that can fit on each panel, the cheaper each PCB will be	~$0.05–$0.12 per square inch for PCB depending on type of board (FR4, flex, etc.), amount of copper, number of through holes, and board utilization among some key cost drivers
Scrap	The scrap for SMT process is usually very low depending on complexity of the PCBA However, the cost of the scrap in absolute terms could be high because the cost of electronic components could be very high	~0.5%–2% of material and process
	Also, the part of PCB panel that is not used as part of the finished PCB will be scrapped	~5%–20% of PCB panel
SMT	This process is almost fully automated with the cost of each line between $1.5 and $2.5 M depending on the number of robotic placement machines needed Usually, at least two or three operators are needed to maintain the SMT line, which includes replacing the tape reels that separately hold each type of electronic component The cycle time per PCBA is determined by the number of electronic components placed, sizes of those components, and number of robot placement machines Each placement is only half a second or less, so a board with 200 electronic components will take total of 15–30 sec to populate	~$0.005–$0.02 per component placement
Tooling	Usually there is very little tooling required since the SMT equipment is designed to process standard panel sizes The only tooling and/or fixtures that will be required are for post-SMT operations	~$20–$60k for all fixtures depending on process

ELECTRONIC COMPONENT MANUFACTURING

There are very many types of electronic components, usually categorized into active (e.g., microprocessors, MOSFETs, power supplies) and passive types (e.g., resistors, capacitors, inductors). Although all of these devices are for use in electronics, they are very mechanical in design, meaning that they are manufactured out of mechanical components and materials. They are also assemblies, so many different raw materials and manufacturing processes are employed prior to final assembly.

"Should cost" estimating concepts have not been heavily employed in the electronic components industry. The prevalent form of estimating component pricing is using historical data with heavy influence of price/volume curves. This means that the price at which customers buy electronic components is heavily dependent on the volume purchased annually. The actual cost to make the components is usually not relevant in purchasing decisions. However, the IC Knowledge LLC integrated circuit (IC) costing software can be useful in establishing base pricing to help customers in negotiations with suppliers.

CUSTOM ASSEMBLY

Whether a part is casted, stamped, molded, or manufactured using some other basic process, it will most likely end up assembled together with other components into an assembly. For example, the brake pads that are used on every car and truck are assembled into a caliper, which is then assembled into a chassis, which is then assembled into a car or truck. In fact, the brake pad itself can be considered an assembly because it consists of a stamped metal plate with friction material molded to it, then machined to size, surface treated, and finally an insulator and pad wear indicator (PWI) assembled to it.

That is a long list of assembly processes and all of them are custom (as in the example in Figure 4.24), meaning that they are customized to the product and cannot be shared with another type of product (e.g., cannot use brake pad assembly line with phone assembly line). Furthermore, there could be multiple ways to assemble the same type of part (e.g., different companies can design completely different and unique brake pad assembly lines). A manufacturing engineer may also choose different levels of automation depending on local labor cost or restrictions on capital investment funds. The annual volume and part design will additionally affect assembly process choices. For low-volume assemblies, a more manual process will most likely be chosen because assembly speed is probably not important.

An assembly process could involve thousands or millions of dollars in investment, two or hundreds of assembly steps, and seconds or hours of assembly time. Due to this variance and complexity, a cost estimator would be advised to view in

FIGURE 4.24 Custom assembly line. (Courtesy of Product by Process; http://www. productbyprocess.com/2018/02/the-amazing-automated-manufacturing-assembly-line-of-the-steam-controller/.)

person as many assembly processes as possible, so that similarities and patterns can be applied in the estimates. What if it is not possible to see a specific process in person? In that case, the cost estimator will have to design the assembly process using simulated motion studies and historical data. Even in those cases, however, it is usually required to have some prior onsite experience of assembly processes to draw from, so that a decent starting point can be developed.

ADDITIVE MANUFACTURING (3D PRINTING)

This is a fairly new form of manufacturing that involves layering metal or plastic material into part shape using heat. Although additive manufacturing has historically been a very slow process relegated mostly to prototype and tool manufacturing, there have been rapid advancements in this technology. For example, additive manufacturing company Carbon uses a combination of light and oxygen to rapidly produce end-use parts from liquid resin for high-volume production (see polymer printing of a shoe sole by Carbon, Inc. in Figure 4.25). While Carbon is already producing parts in the scale of millions, it is expected that additive manufacturing technologies on the whole will need another 5–10 years of development to be employed across the full spectrum of materials and allow for widespread high-volume production. It is possible that in time all previously described manufacturing processes will be eliminated

FIGURE 4.25   Additive manufacturing. (Courtesy of Carbon, Inc.; https://www.carbon3d. com/our-technology/.)

and replaced by additive manufacturing. This will allow for greater flexibility in part design and perhaps less capital, energy, and labor investment. It will also allow for shorter design and re-design times since no tools would be required to be built and design changes would only require changes in CAD (computer-aided design) models. Since developments in this manufacturing process are very rapid, the cost implications will not be discussed in this book.

SUMMARY

Cost estimators sometimes spend years learning all the intricacies of the part designs and manufacturing processes that they are estimating. This book does not aim to capture all of those intricacies, but instead gives a basic overview of all the most commonly used manufacturing processes and main cost impacts that need to be considered. Those embarking on cost based negotiation, either buyers, engineers, or cost estimators, might need to reference these basic concepts in order to be successful.

Although processes such as molding, stamping, casting, and machining are staples of the manufacturing world today, there are many others that could define the world of tomorrow. The additive manufacturing process could be a wave of the future that gives engineers a very different set of design and manufacturing options. This technology is still developing and not used much for high volume production, but it will be important to understand the costs associated with it when it becomes more mainstream.

5 Cost Allocation Methods

The lack of accurate cost allocation method is one of the biggest reasons why companies are unsuccessful. This is because inaccurately allocating cost will lead to either under- or overpriced products, one leading to selling products that do not make money and the other leading to not selling products at all. While sometimes these could be overcome with pure luck or having a product that has such high demand that a company can charge much higher prices than required, this is rarely the case and most companies must operate in very competitive environments with relatively low margins.

It is also true that most companies are not aware of whether their allocation method is accurate or not. If a company does not know that is has a cost allocation problem, it will not seek a solution to that problem. Most go about their business assuming that their allocation method is accurate and blame other things for lack of profit or lack of business.

If you are reading this book and the only thing you get out of it is that you possibly might have a cost allocation problem, then you already have put yourself on a path to success. At least now you can seek help. However, before we get into the right way to allocate cost, let's first discuss some of the most popular allocation methods, starting with one of the least accurate ones.

CONTRIBUTION MARGIN AND GROSS MARGIN ALLOCATION METHODS

These are by far the most inaccurate allocation methods because they simply assign the same margin percentage to all products and usually to cost that a company finds easiest to measure such as material, labor, and maybe some other minor items. The distinction between contribution and gross margin is that one is a difference between price and all variable costs while the other is a difference between price and all manufacturing cost.

The margin, either contribution or gross, is usually something that a company is trying to achieve for its stakeholders and is tracked based on historical data. For example, a company might analyze last year's finances and find that its gross margin was 20%, meaning that all its manufacturing cost was 80% of revenue and the remaining corporate cost and profit was 20%. Since this company considers corporate overhead as fixed (not changing with time or product volume) and it does not know how to assign its corporate overhead to products, but wants to improve profit, it might simply target to improve its gross margin to something above 20%. As it quotes new business, regardless of what product it tries to sell, the cost estimating group will apply this higher margin on top of the calculated manufacturing cost. In this way, it hopes to achieve improvement in its profit over time.

Why is this method inaccurate? It is because the contribution and gross margins, in absolute terms, will be different for different products and different product volumes. A product with low engineering requirements and high volume requires smaller margin percentage than a product that is highly engineered, requires a lot of customer interface, and/or has low product volumes. By applying the same margin percentage to every product, a company will be penalizing some products and underestimating others. Over time, this will result in company winning business that is underestimated and losing business that is overestimated, which ultimately results in less profit and perhaps bankruptcy.

REAL-LIFE ANECDOTE

A Tier 1 supplier of electronic components was using a flat 10% assumption for SG&A cost, meaning that regardless of what type of business was being quoted a 10% was added to the piece price. This 10% was based on historical SG&A cost for this company. Up to this point, the company was winning some smaller business but was uncompetitive on larger business even though the company was very lean and agile. However, during its quote process for a business that would double the size of the company, the new quote manager pointed out that the company's SG&A cost in absolute terms would not double but only increase by about 25%, thus a lower SG&A percentage should be quoted. With this assumption, the company was able to submit a quote several percentage points lower than was typical and ended up winning the large business.

LABOR ALLOCATION METHOD

The method of allocating cost, especially overhead, based on a number of operators or the cost of labor is a very simple method, but can also be very inaccurate. For example, based on historical data, such as previous year's profit & loss (P&L) statement, a plant controller (chief accountant) can determine that if $1M was spent on labor and $5M on overhead, then going forward, every time product cost is estimated, for every dollar of labor needed, five dollars of overhead will also be required.

Sounds like a reasonable method of allocating cost, correct? Unfortunately, it only sounds reasonable to a lot of people because it is one of the most popular methods of allocating cost in manufacturing (at least in the United States). Its simplicity is what is attractive and perhaps for simple, static manufacturing operations it could be accurate enough. Where this method fails is in cases where manufacturing operations are complex and dynamic.

For example, if the manufacturing plant in question only makes one casting or several castings that are very similar using the same casting process, then the ratio of labor to overhead would always be the same. However, what if the casting facility had three casting machines of different sizes and the product mix varied from simple parts to complex ones, each needing a different amount of labor to process it? Would a casting requiring four operators employ less overhead cost than a product requiring eight operators while running on the same casting line? Of course not. The machine and the cost to run it are the same regardless of how many operators are needed.

Another example is a company that decides to add more value for their customers so it adds assembly to its stamping operations. It also determines to use the same 10 to 1 ratio of overhead to labor that it has seen historically in its stamping operations. As the company quotes new assembly business, which includes 10 assemblers, it adds $10 of overhead for every $1 of labor cost. The company is surprised when it does not win that business and the next 20 assembly businesses that it quotes. The company finally gives up assuming that its cost is simply uncompetitive for assembly processes.

What happened, in actuality, was that the company overstated its overhead cost for assembly. While it may be true that 10 to 1 ratio is accurate enough for its stamping operation where one operator runs large presses that require large capital investment and lots of space and energy to operate, the assembly operation only requires inexpensive table-top stations that require little space and energy to operate. Thus, the overhead cost to run an assembly process is much lower than stamping. The real ratio might actually be only 4 to 1 for an assembly operation; so the company was dramatically overpricing their assembly operation and could have perhaps won some profitable business if an accurate cost allocation was used.

In summary, the labor allocation method is not recommended for most companies as it can be very inaccurate, unless the manufacturing operation is only for one product or very similar products using the exact same process. Even in those cases, however, it is important for the person responsible for cost estimation and allocation in such a company to be aware of its pitfalls and be able to adjust with changes in the company's operations.

REAL-LIFE ANECDOTE

Tier 1 supplier of electronic control units (ECUs) had a manufacturing plant that was producing for both automotive and appliance industries. The process flow to manufacture products for both industries was very similar as it consisted of SMT and Final Assembly. The company felt justified in using the same allocation of overhead based on labor cost for both types of product. However, it was not considering the down time as part of the labor cost and the down time was dramatically higher on appliance business because it was low volume and high product mix. Operators spent a lot of time changing over to different products or just sitting around waiting. Thus, the overhead cost was under allocated for the appliance business and the automotive business was absorbing that overhead. Ultimately, whenever the company was quoting business, it was underpricing the appliance parts and overpricing the automotive parts. Also, the company's management was making decisions with an understanding that their appliance business was making a lot of money and their automotive business was underperforming, when in fact the opposite was true.

STANDARD HOUR ALLOCATION METHOD

Similar to the labor allocation method, the standard hour allocation can be problematic. In this method, a manufacturing company assigns a target cycle time, or standard, needed for every process, so that it can measure variances to that standard, which gives it an ability to flag and address production problems. In a similar manner, the company's accounting department allocates cost to these standard times, on a per-hour basis, using the previous year's P&L results. For example, if a company has an overall operation cost of $20 million and between all its processes there are 500,000 total hours of machine time (e.g., 100 machines each running 100 hours per week and 50 weeks per year), then the cost per standard hour of machine time is $20. Now, if a machine is out of service or runs slower than usual, then the accounting department can say that the plant cost is running higher than what it is supposed to due to increased cost, versus standard, for that machine. They report these differences as variances to standards.

The same cost of $20 per standard hour of machine time is used for all future customer quotes, which might be fairly accurate if all 100 machines are the same and make very similar products with the same process. However, what happens if there is variation in the machines or the manufacturing process? What if a part requires two operators instead of one to handle the manufacturing process? Or, what if the needed machine is more expensive or requires more energy or maintenance? The company would still use the same $20 per standard hour. This would possibly underprice the product and most likely lead to the company winning the business, and, unknowingly, lose money on that business.

Over time, the company's overall cost will go up, but all it would do is take the increased overall cost and amortize it over total annual standard hours of operation. Most likely, the company's rate per standard hour would go up for all products. This ultimately leads to the company being less competitive in the marketplace, perhaps losing business, thus reducing utilization of those 100 machines, spiraling into further increases in hourly cost per standard hour (because accounting usually only uses last year's P&L totals to spread over standard production hours). This, in turn, leads to losing even more business, and eventually the company having to close down the manufacturing plant.

It is important to keep in mind this flaw of standard hour allocating method when quoting products. As a cost estimator, it is important to be able to use common sense and adjust cost based on products being quoted and not to be restricted by rigid cost accounting rules. Otherwise, the company will suffer from its own ignorance and never even know what happened.

CAPITAL ALLOCATION METHOD

Another method similar to the labor allocation method is a method of allocating cost based on capital employed for each process and developing hourly rates for each machine based on its original acquisition cost. For example, if the total overhead cost for a factory is $10 million and there are five machines, two that were bought new for $200k each, one for $600k, and two for $500k, then the $10 million will be allocated as in Table 5.1.

TABLE 5.1

Capital Cost Allocation Method Example

Machine	Machine Original Acquisition Cost	% of Total Machine Cost	Cost Allocation per Machine	Hourly Rate per Machine (Assuming 5,200 Annual Hours)
1	$200,000	10%	$1,000,000	$192.31
2	$200,000	10%	$1,000,000	$192.31
3	$500,000	25%	$2,500,000	$480.77
4	$500,000	25%	$2,500,000	$480.77
5	$600,000	30%	$3,000,000	$576.92
TOTAL	$2,000,000	100%	$10,000,000	

After hourly rates are calculated for each machine, an estimator will multiply the cycle times for each individual part being quoted by these hourly machine rates to get the overhead cost per part.

As with the labor allocation method, the capital allocation method can sometimes work for very consistent and static manufacturing operations, but is mostly inaccurate and outright dangerous to the well-being of a company in all other cases. What if, in our earlier example, one of the two $500k machines was actually an assembly line? Would its overhead cost be the same as the other $500k machine that is perhaps a molding press? Probably not. An assembly line might require more people to support or take up more space, but it will usually require less electricity and less maintenance. The only thing that is the same, in fact, is the original cost of the equipment.

MACHINE RATES

Similar to the capital allocation method, machine rate methodology defines rates for each machine in the manufacturing plant, but the allocation itself tries to be more accurate by assigning cost that is more realistic and specific to each particular machine. This means defining rent cost for the space that a machine is using, the electricity that it uses, the maintenance and supplies that it requires, etc.

Unfortunately, not all cost is easily assigned to a machine, so some of the fixed overhead cost (cost that does not change with product volume or with time), such as the cost of plant manager salaries, is still allocated based on some other, probably less accurate basis. Nevertheless, this method is still more accurate than all the other previously mentioned methods. Table 5.2 shows an example of how cost might be allocated for the same five-machine factory that we described earlier.

Note that the hourly plant overhead is much higher than all the other cost categories and those rates are derived by multiplying all the other categories combined by a crude 200%, meaning that costs that the cost accountant or cost estimator is not able to allocate directly to the machines – such as plant manager, accountants, HR, IT, office space, parking lot, office utilities, outside services such as cafeteria or buses or uniform washing, etc. – are determined to be two times that of all the costs allocated

TABLE 5.2
Machine Rate Cost Allocation Example

Machine	Machine Original Acquisition Cost	Depreciation Hourly Cost Assuming 5200 Annual Production Hours and 10-Year Schedule	Hourly Space Rental Cost	Hourly Electricity Cost	Hourly Indirect Support Cost	Hourly Maintenance and Supplies	Hourly Plant Overhead (200% of Other Mfg Cost)	Hourly Rate per Machine	Total Annual Cost
1	$200,000	3.85	22.00	20.00	16.00	20.00	163.69	245.54	$1,276,800
2	$200,000	3.85	22.00	20.00	16.00	20.00	163.69	245.54	$1,276,800
3	$500,000	9.62	39.00	40.00	16.00	40.00	289.23	433.85	$2,256,000
4	$500,000	9.62	39.00	40.00	16.00	40.00	289.23	433.85	$2,256,000
5	$600,000	11.54	52.00	50.00	16.00	60.00	379.08	568.62	$2,956,800
TOTAL	$2,000,000								$10,022,400

to the machines, per the P&L statement from the previous year. Although the effect is exaggerated in our example, the inaccuracy of using such an allocation could be very great. For example, what if most of the non-machine allocated resources are dedicated to products that run on the $200k machines? If that is the case, then the machine rate allocation method would severely underestimate the cost of running on those $200k machines and overestimate cost on the other machines.

What has not been discussed yet, but could also be a significant cost is the corporate overhead such as sales, finance, purchasing, and executive management. These are not part of manufacturing cost and not easily allocated to any specific product. This cost is usually added on top of all the other cost as a straight percentage (e.g., 10% added on top of all other cost) under an assumption that all products will incur the same corporate overhead cost on a percentage basis regardless of product type, volume, manufacturing process, or customer. This increases the chance of inaccuracy because excessive re-quotes, engineering changes, and "freebies" demanded with the product or customer behavior will generate a higher customer service add-on percentage for some customers.

Although the machine rate methodology has its own potential pitfalls, it is becoming one of the most popular methods of cost allocation, because it offers decent accuracy with affordable effort to calculate. As long as most of the cost can be allocated directly to the machines, this method offers better accuracy than the previously described methods. However, it is always important to use common sense and make sure that proper adjustments are made in cases where rules might not apply.

ACTIVITY-BASED COSTING (ABC)

The most accurate method of allocating cost is the ABC method. This method defines activities that happen in a company and then allocates cost to those activities using different causality drivers (things that cause cost to happen). These activities could be manufacturing processes themselves, but will also include others, such as selling of products, processing of accounting transactions, or movement of raw materials inside the factory. Note that activities not related to manufacturing would try to address the allocation of cost that is not easily allocated to production machines. This is so no cost, or very little, is left to be arbitrarily spread over products like peanut butter over toast as is done in other cost allocation methods. Figure 5.1 shows a simple diagram describing the ABC allocation method per Douglas T. Hicks.

There are many books on ABC, therefore we will not go into detail here. However, because most of the books are only conceptual in nature, the recommended one for its practical examples, detailed explanations, and use of common sense is "Activity Based Costing: Making It Work for Small and Mid-Sized Companies" by Douglas T. Hicks. Although the title indicates usage for small- and mid-sized companies, the methodology can also be applied for large companies.

It is important to note that ABC was embraced by many companies in the past, but has recently fallen out of favor and been discontinued by some due to its intensity of time and resources. Although this intensity is not necessary, some companies began to measure every possible activity that they could identify, which led to excessive effort for very little return. This extreme detail and resulting effort are not necessary

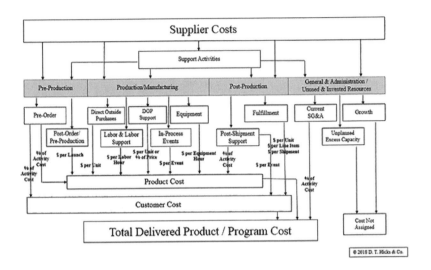

FIGURE 5.1   ABC allocation. (Courtesy of Douglas T. Hicks.)

to develop an accurate and predictable cost model. As long as the key activities are defined and drivers are properly assigned, then the cost model can generate good enough accuracy. It is failing to capture the correct activities or drivers that have a much higher chance of inaccuracy. As Mr. Hicks often suggests, accuracy is much more important than precision. Being accurate means that you are hitting very close to the bull's eye, while precision means that you are hitting exactly the same spot every time, except that the spot may not be anywhere close to the bull's eye.

SUMMARY

There are many methods of allocating cost, but the most important thing to consider in allocating cost is using common sense. There is no one size fits all, so while one method works for one company, it might not work at all for another. As a cost estimator, one must follow cost logically instead of locking himself/herself into a box of specific rules and methodologies. Do not get bound by past cost and instead try to predict what the cost will be in the future state that the company seeks, e.g. with optimal machine utilization.

Cost allocation is the most important building block of product cost estimating, so getting it right is critical to achieving good accuracy of estimates. Otherwise, no matter how good the rest of your costing methodologies might be, you will be way off from the bull's eye. As navigators of your company's future, it is up to you to use the proper navigational tools.

6 Cost Optimization Tools

Now that we have established the proper cost estimating principles, it is time to discuss how to use those and other tools in terms of cost optimization. The first step is prioritizing the cost optimization initiatives based on the data that is now available through cost estimating efforts. There are many companies that spend an inordinate amount of time optimizing the cost of things that contribute only a small percentage to the overall cost, because they have no idea what their cost is and how to prioritize it.

One example that comes to mind is that of companies that are mostly assemblers of products. These companies purchase all the components that go into an assembly, then assemble those components into a complete finished assembly (e.g., automotive chassis). Companies like these often invest a lot of time and resources to optimize their assembly factories, which only account for 5%–10% of overall cost, but completely ignore the suppliers of the purchased components, which often account for up to 80% of the overall cost. If the investment in resources assured an improvement of 10%, would it make more sense to invest in 10% of cost or would it be better to invest in 80% of cost, meaning would it be preferable to save 1% or 8% on total cost?

A company will sometimes make a realization that it needs to invest in helping its suppliers reduce cost, but then it will only go as far as a "lean event" at their factory, where only the supplier's manufacturing process is in focus and completely ignores the rest of the value chain, even if that supplier is only an assembler itself. The key, therefore, is to first understand what the cost is and then be able to prioritize the cost optimization efforts by the largest cost.

With prioritization established, the next step is to employ appropriate cost optimization tools. The cost engineer has many tools and techniques to choose from based on a type of cost that needs to be addressed. For example, lean might be employed when manufacturing process is an issue, value stream mapping (VSM) when the full supply chain needs some work, or functional analysis and/or TRIZ combined with benchmarking for design issues. These techniques are not the only ones but are the most commonly used. The next sections provide a brief description of each.

LEAN MANUFACTURING

One of the most well-known and used manufacturing optimization tools, lean focuses on making the manufacturing of a product as efficient as possible. This means manufacturing a product in as little time as possible with as few resources as possible. The three key items of lean optimization are cycle time, overall equipment effectiveness (OEE), and inventory. Optimization initiatives for the first two items focus on optimizing the time it takes to produce parts, while inventory initiatives focus on reducing inventory levels (resources) throughout the value chain. This includes the factory's incoming, in-process, and outgoing inventory, but often extends to the inventory in between customer and supplier factories.

The lean initiative mostly focuses on the factory itself but can extend to the full supply network (e.g., value stream mapping). In terms of using lean as a cost optimization tool, it is mostly useful for companies that have not embraced lean methodology already. This is where the biggest opportunities can be found. For companies that have been using lean methodology for an extended period of time, this tool might not lead to significant savings. For example, if a machining company that has used lean manufacturing finds a 10 seconds cycle time reduction opportunity for a product that requires 500 seconds in machining, the lean team might celebrate that as success, but if no laborers are eliminated or no new product is machined in those saved 10 seconds, then there will be no real cost savings. The savings might only materialize over a period of time after incremental improvements lead to enough open capacity that it can be utilized for more products that absorb more of the fixed and some of the variable costs.

Regarding inventory, the goal is to minimize it to the lowest possible level since inventory is considered to be just cash sitting on the shelf. The kanban concept is often employed to have only minimal levels of inventory with just-in-time (JIT) as the best-case scenario where the factory only builds what the customer orders that day. It is important to note, however, that extreme focus on lean can create some issues down the stream in the supply chain. For example, if the customer factory is extremely lean, but lean is not extended to the supply chain, then those suppliers could be burdened with carrying excess inventory that the customer did not want at their factory. If suppliers are carrying one month's worth of inventory to allow for its customer to be lean, the suppliers will eventually pass on the cost to the customer and the cost savings will be minimal or non-existent.

There are hundreds of books available on lean manufacturing for detailed understanding. Two books with good basic descriptions are "Lean Thinking: Banish Waste and Create Wealth in Your Corporation" by James P. Womack and Daniel T. Jones and "The Toyota Way: 14 Management Principles from the World's Greatest Manufacturer" by Jeffrey Liker.

VALUE STREAM MAPPING

A good start to a cost optimization effort is a value stream map (example shown in Figure 6.1), which describes the movement of material from beginning to end of the product manufacturing process. It starts with a customer sending an electronic order for parts, which alerts suppliers of raw materials to send more material to the factory for processing and assembly, and ends with the finished product being delivered to the end customer. The VSM in Figure 6.1 is a simplified map where only one supplier of raw material is involved and only one manufacturing facility is finishing the product. In reality, especially for assemblers, there could be multiple raw material and purchased component suppliers and multiple factories producing parts with all of them later being sent to the assembly plant before packing and delivery to the end customer.

A typical VSM tells a story of how the finished product was produced, but also describes each manufacturing process as far as number of operators, cycle times, uptime, inventory levels at each step, time it takes to ship, and places that the material

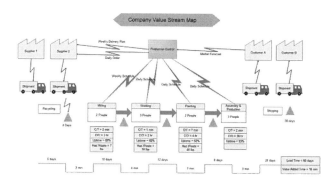

FIGURE 6.1   Value stream map. (Courtesy of latestquality.com; https://www.latestquality. com/create-value-stream-map.)

travels to throughout the manufacturing process. Having this information is a great cost optimization idea generator since it makes any inefficiency very obvious. It is very likely that a company that has never developed a VSM will find glaring opportunities during this process such as elimination of multiple transport of the same part, sometimes across the ocean, or reduction of excessive inventory across the value chain, or elimination of bottlenecks that prevent optimal equipment utilization.

One example of value chain inefficiency is an automotive OEM that received a corporate direction in the 1990s to have 60% of its purchased components sourced out of China. This generated an extended supply chain with cheaper parts bought in China, but much higher logistics and inventory costs. What also increased was the cost of scrap, because of the lower quality China parts, and the cost of supporting personnel that had to spend a significant portion of their time developing and managing suppliers in China. This significantly increased the overall cost of parts for this OEM.

Overextended supply chains and other inefficiencies can be uncovered by a value stream map, but its greatest benefit might be the ability to visualize and then prioritize cost optimization efforts. For example, if the total cost of an assembly is $100, but only a few purchased components contribute more than 50% of the overall cost, then these parts should be prioritized in the cost optimization effort. Focusing on the top cost drivers will give a company the biggest bang for the buck.

"Value Stream Mapping: How to Visualize Work and Align Leadership for Organizational Transformation" by Karen Martin and Mike Osterling and "Value Stream Mapping for the Process Industries: Creating a Roadmap for Lean Transformation" by Peter L. King and Jennifer S. King are recommended for detailed explanation and practical examples of VSM.

BENCHMARKING/TEARDOWN

Another effective cost optimization idea brainstorming tool is tearing down the competitors' products and comparing them to your own on a cost basis. This activity is called benchmarking, and it is most effective when cost estimates are developed for every component individually (see the example in Table 6.1). Not only does this

TABLE 6.1
Benchmarking Cost Analysis Example

	Our Part	Competitor #1	Competitor #2	Competitor #3	Competitor Best Practice
Component #1	$2.56	$2.13	$2.00	$3.50	Zinc coating instead of silver
Component #2	$5.43	$7.00	$4.78	$3.73	Lower casting weights
Component #3	$10.44	$10.22	$9.55	$9.40	Less copper and smaller magnets

help prioritize, it also gives those involved in benchmarking an idea of what things cost in general. Oftentimes, the people asked to participate in benchmarking activities have zero or limited costing background, so having all the components costed up front avoids generating ideas that are actually cost increases (e.g., using competitor's style seal that actually cost more than the seal used in your assembly).

Some engineers might question why benchmarking is a good idea. After all, their designs are superior, so why would anyone want to look at the competitors' products? This feeling is not uncommon because engineers spend long hours designing products and grow attached to those designs. Engineers also focus primarily on part performance and prefer not to touch something that has proven to work in the past.

These types of attitudes often make benchmarking exercises ineffective, so it is important to have this activity driven by the top management and supported by well-trained facilitators who know how to motivate a team to find and implement cost reduction ideas.

FUNCTIONAL ANALYSIS

One of the most powerful design cost optimization tools is the functional analysis under the umbrella of value analysis/value engineering (VA/VE) methodology. It helps to generate cost reduction ideas by evaluating every function that is performed by the product and asking if those functions can be achieved at a lower cost or if the function can be increased at the same cost. Either one of those solutions would increase value to the customer with value defined as function divided by cost.

As a simple example, the main function of cars is to provide people with an ability to get from point A to point B. This function is currently mostly achieved using gas powered engines, which has obviously provided humanity with great value, but using functional analysis one might ask, can humanity get the same or greater value by achieving the same or better function at a lower cost? This is exactly how alternative powertrains have been born with the electric ones currently achieving most success. The electric cars claim to offer better functionality – such as less noise, faster acceleration, and lower carbon footprint – with cost that is approaching that of conventional gasoline engines.

Functional analysis is also effective at the micro level. Using the same electric powertrain example, every function of the powertrain can be analyzed. For example, asking the question if the same or greater battery life (better function) can be achieved at a smaller size (lower cost), solutions such as different raw materials and optimized electrical routings were developed. By continuously asking such questions, engineers knowingly or unknowingly have been using functional analysis to evolve the battery design over the last couple of decades from very large ones that could hold a charge for only a few miles to batteries half that size that hold four times the charge while decreasing the cost of batteries tenfold.

There are many books on VA/VE methodology, starting with "Techniques of Value Analysis and Engineering" by Lawrence D. Miles, who is considered the founder of this methodology. VA/VE is also fostered by a professional organization called SAVE International that actively markets these methods in many industries such as civil engineering, aerospace, defense, automotive, and many others.

TRIZ

During the functional analysis exercise, a team often finds itself struggling for alternatives to providing a function at lower cost. This is typically the biggest roadblock in the brainstorming activity. Fortunately, a Soviet-era scientist and inventor, Genrich Altshuller, developed a method called TRIZ, which in Russian stands for "theory of the resolution of invention related tasks." Altshuller analyzed thousands of patents and came up with 40 most common solutions or patterns of resolution to most typical problems (see Figure 6.2). These then can guide a person or a team throughout the creative design process.

For example, one of the results of using the "converting harm into benefit" solution (#22 in the chart) is using a braking system to charge the battery in an electric vehicle. Another example is "using electromagnetic system as replacement to mechanical one" (#28 in the chart) as in a motor replacing an engine in automotive vehicles.

There are several books on TRIZ, such as "And Suddenly the Inventor Appeared" by Altshuller himself, and there is a lot of information about TRIZ online as well as some apps with excellent visual aids to guide the user.

COST REDUCTION WORKSHOP

A very common tool in the cost optimization tool belt is the cost reduction workshop. This is where several people are pulled together for a length of time to brainstorm cost reduction ideas. Unfortunately, these workshops are often poorly organized and misguided due to multiple factors. In order for the workshop to be successful, several key ingredients must be in place. These include obtaining support from top leadership and dedication of the attendees, attendance of the right people with cross-functional product knowledge, establishing challenging and motivating cost reduction targets, having a proper process and a skilled process facilitator, using the right tools to open the minds of those attending, having the right facilities, and giving it enough time (3–5 days). SAVE International has established a great set of guidelines, but those sometimes can be restrictive in the process and tools that are used.

FIGURE 6.2 TRIZ 40 principles. (Courtesy of FotoSceptyk and Wikimedia Commons; https://commons.wikimedia.org/wiki/File:40_principles_of_TRIZ_method_225dpi.jpg.)

For example, a facilitator might insist on using functional analysis combined with Alex Osborne brainstorming technique, but teardown analysis with TRIZ technique might be more appropriate. It is important to maintain flexibility, especially when the workshop concept is new to a company.

The most critical element of the cost reduction workshop is harnessing the team's creative power by bringing different perspectives to the same problem. This can be accomplished by engaging the cross-functional members of the organization or by reaching outside the company to consultants. At the same time, it is also important to harness the power of ignorance by bringing into the workshop those who might lack product knowledge and can ask the "dumb" questions that spark ideas. Ignorance is bliss but it can also be very useful in cost optimization. These elements will allow for creativity and innovation to flow without which success will be limited.

REAL-LIFE ANECDOTE

An automotive OEM made a drastic improvement in cost saving results when the company's CEO gave a directive to establish cross-functional teams for each of the components/commodities that it was buying from suppliers. He also gave them a 30% cost reduction objective to be achieved in the next five years for their respective spend. In addition, every function was given performance objectives to meet the same objective. Faced with this challenge, the teams had no choice but to rally together to come up with a strategy that spanned technology improvements, design changes, commercial negotiations, logistics improvements, and many others. Instead of meeting occasionally, the teams were co-located for three months outside of their normal office space, virtually turning it into a very long workshop. The result of this initiative was a drastic decrease in cost with most teams reaching their 30% reduction targets.

CONNECTOR EXAMPLE

Cost optimization tools can be very useful in improving the cost of our connector. After the cost engineer brings a cross-functional team together for a cost reduction brainstorming workshop and goes through various exercises using the previously described tools, the following opportunities are found:

- Per the connector's value stream mapping exercise, the team finds that although the final assembly takes place in Mexico, the pins are stamped in California, the wires are made in China, and resin comes from Texas. This discovery sparks ideas to localize all three in Mexico which should, at a minimum, provide logistics and inventory savings.
- Per the benchmarking evaluation, the team discovers that their competition uses nickel plating on pins instead of silver plating, which is a large cost saving opportunity. It also finds that the resin used by others probably has less glass fibers (GF15 vs. current GF30), which would also provide some savings. This sparks another idea to allow for re-grind and re-use of scrapped resin during molding, which provides even more savings.

- Finally, during analysis of connector functions and their cost, the team discovers that one of the wires can be eliminated. In fact, using a TRIZ brainstorming session, the team realizes that all the connector functions can be replaced by wireless communication. However, the team decides that this idea is not feasible at this time.

Most of the generated cost reduction ideas are not drastic and would normally be obvious. Nevertheless, these opportunities were missed or forgotten as the team rushed to develop and launch a working product. Since product performance always takes priority, members of a cross-functional team often do not have time to think about cost optimization. This is why taking the time to focus specifically on cost optimization can lead to significant impact on profit.

SUMMARY

When cost is well understood, a company has a much better chance to prioritize its cost optimization efforts. It can then use a set of tools for those efforts, such as Value Stream Mapping, benchmarking/teardown, functional analysis (VA/VE), and TRIZ. These are designed to open the eyes of the cost optimization team to potential opportunities either in value stream, versus the competition, or in how functions are performed by a given product.

A workshop is often the most potent cost optimization tool because it dedicates a cross functional team for multiple days to the effort. In this time, the team can analyze the data related to a specific design and use all the other cost optimization tools at their disposal. The cost reduction workshop, however, must be organized and facilitated properly in order to get the most out of the people and time dedicated to the initiative.

7 Optimal Cost Engineering Process

As we saw in the SPARK hybrid truck example, the lack of cost engineering focus during the initial vehicle design stage resulted in cost overruns that seemed impossible to resolve. After 2 years of engineering development, the vehicle was already 45% more costly than permitted ($130,000 versus target of $90,000). There also seemed to be no solution on how to lower the cost because (1) the Engineering team was already too far advanced with their designs and too afraid to compromise the vehicle performance with any redesigns and (2) the Purchasing team had no leverage to lower supplier costs since the sourced volumes were relatively low and had a lot of risks associated with it considering the vehicle concept was new and SPARK has never sold a single vehicle.

What happens next can be viewed as critical to the viability of a company. Unfortunately, the SPARK hybrid truck company has only two paths to choose from. The first path is to spend an additional year or two redesigning the vehicle. This path could prove to be fatal since it will cost another $10,000,000 per year (100 engineers × $100,000 average salary) and the company will probably run out of cash before it achieves an acceptable redesign. Also, since the company still doesn't know what each individual vehicle component should cost, there is a good likelihood that it will arrive 1 or 2 years later with a design that is still too costly. Finally, without proper cost targets, engineering will most likely just de-content the vehicle (eliminate functions), which would most likely mean compromising on performance and ultimately loss of vehicle sales.

The second path would be to make a tough choice of increasing the vehicle price to above the market (e.g., to $150,000 per vehicle). However, since the company already analyzed the market pricing and found that the price would have to be around $100,000 in order to sell 10,000 vehicles annually, this would mean that any price above that would negatively impact vehicle sales. Fewer vehicles would have to cover the same amount of in-house expenses for both SPARK and its suppliers. This would further increase the cost per vehicle and most likely reduce or eliminate any profit.

Both paths likely lead to the same result: the company losing money on their product and possibly going bankrupt and having to lay off those 100 engineers as well as hundreds more from other functions within the organization. Perhaps even worse, the consumers would unnecessarily lose a chance to purchase a product that could have been superior in function and value to other alternatives.

COST ENGINEERING PROCESS

What, then, is the proper cost engineering process and when to use each cost optimization tool? It is very common for companies to try and cost optimize at random and often at the wrong times in a product life cycle. For example, some automotive OEMs have had a bad habit of organizing cost reduction workshops only a couple of months before the start of production, after designs are already finalized and tools are already built. Such a workshop would normally be kicked off by the vehicle line executive with a statement like this: "We are 20% over on our vehicle's cost, so we need your ideas on how to reduce cost so that we can make this vehicle profitable." Obviously, it is too late in the development process to ask for such cost reductions. The vehicle is about to go into production and any design changes would require significant time and money to validate and re-tool. It would be akin to building a house, then deciding to completely remodel it because it is over budget.

It is important to have a cost optimization cadence throughout the product life cycle. The term VA/VE, which stands for value analysis/value engineering, is already trying to give a hint of this, because value engineering is defined as all cost optimization activity prior to design freeze or production start, while value analysis is all cost optimization activity after design freeze or production start. However, it is most effective when cost optimization is built into the company's product development process with checkpoints and activities at every product development stage as in Figure 7.1.

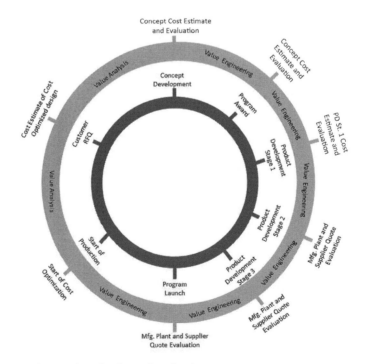

FIGURE 7.1   Cost engineering in product development process.

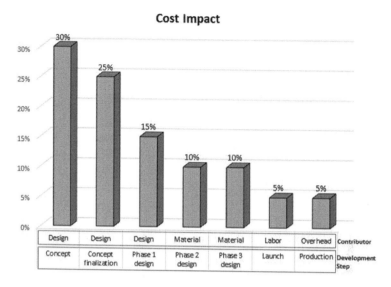

FIGURE 7.2 Cost impact by each stage of product development cycle.

In addition, Figure 7.2 shows the influence each stage has on the ultimate cost of the product along the concept to production continuum. It is important to note that the impact on cost is biggest early in the process, during the design development process. The earlier in the product development a company can focus on cost optimization efforts, the better are its chances of having a profitable product.

Some would argue that tying the designers' hands with cost restrictions at the concept and early design stages is a bad idea, because it could prevent them from coming up with the most creative designs. However, designers already have a lot of other restrictions, such as space, weight, and safety, among others. It is rarely a blank sheet of paper that the designers have as their starting point. In fact, in most cases, the product already exists in production and the designers are tasked with updating and improving on that existing design. In reality, the cost restriction is already there, because the cost of the existing product is known.

Even in the "blank sheet of paper" situations, the cost restriction already exists, because designers know roughly what consumers will be willing to pay for a specific functionality of a product. For example, if the car designers were told that their mass market car has to be able to drive itself, they already know that consumers will only pay so much more than they would pay for a regular car. Only very few people would be able to afford the car if, for example, it doubled in price. It would then become a niche product, not a mass market product.

Therefore, it is reasonable to implement the cost restriction early in the product development process. One could even argue that such restriction would actually increase creativity since it forces designers to dig deeper to find solutions to their problems, thus possibly unlocking groundbreaking designs. Also, the alternative is that designers do not have cost restrictions and end up designing a product that costs more than what consumers are willing to pay, which means no sales and maybe no company.

CONCEPT COST ESTIMATE AND EVALUATION

Before engineers put their designs in a computer 3D drawing/model format, they first sketch a rough design concept or markup existing drawings for modified designs. Many engineers have stories about design concepts being no more than a paper napkin sketch. Unfortunately, most engineers also admit that cost is usually not discussed at this stage.

Cost MUST be evaluated at the concept stage. It is absolutely critical to evaluate cost at this point in the process. Even though the design details would be very limited, an experienced cost estimator should be able to provide a ballpark cost of the design. Based on raw materials used, the rough part weight, and manufacturing processes needed to make the part, a cost estimator should be able to tell if the design will cost $1, $10, or $100. Even if the estimate is only within 10% accuracy it would be sufficient to help engineering and the company to make a decision on whether to move forward with the design or not.

In the SPARK hybrid truck example, had engineers been provided with the initial estimates of their designs – especially for brand new designs of components such as motor, battery, and other electronics – they would have had knowledge that their design was exceeding the cost targets early enough in the design development to change course. They would not have to wait for 2 years, after the designs were finalized, to find out that they need a major redesign due to excessive cost.

It is worth noting that some companies do not realize they are over their targets until after the tools are already manufactured or in the process of being manufactured. This occurs because most companies have only a short time to develop products, so they begin tool manufacture after only the initial designs are complete. These initial designs often do not work and must be improved. Engineering must then push redesigns quickly due to timing and pressure to meet performance requirements for the part. As a consequence, these redesigns are often very conservative and ignore any cost implications. They also usually mean significant changes to the tools. The company's suppliers have all the leverage in these situations and are likely to charge premiums both on piece price and tooling.

The company might ultimately arrive at a point where it is very near production, but it finds that its product's cost has risen to the point that it cannot sell it at an acceptable profit. What follows are rushed cost reduction workshops and yet another round of redesigns with very little time and no supplier leverage. At that point, a company will probably wait until after production start to optimize cost, but the cost of additional engineering and tooling redesign is significant and often enough to keep the product from ever becoming profitable.

Real-Life Anecdote

It was a standard practice for a major automotive OEM to hold cost reduction workshops due to significant cost overruns only three months prior to start of production, after all the tools have been built. This happened on almost all vehicle launches because engineers soon realized that cost was not important during development so they adjusted their behavior to concern themselves only with meeting performance requirements. It became company's process and culture to work on fixing cost issues just prior to or after production start even though it cost the company millions of dollars to do so.

PRODUCT DEVELOPMENT COST EVALUATION

After the initial concept is evaluated and the company decides to move forward, the product development process begins. Usually, this is a multi-stage process with several checkpoints, sometimes called gates, along the way to evaluate the design progress. At those checkpoints, companies usually focus primarily on design performance and very little, or not at all, on cost. This is especially true very early in the product development process when designs have not been finalized so supplier and in-house manufacturing engineers have no basis on which to provide cost estimates. Those estimates are usually provided only after the design drawings are finalized.

It remains critical to evaluate cost at each product development checkpoint, and perhaps more often, even when design information is still preliminary. From initial concept to the first development checkpoint, the design could change so drastically that a corrective action must be taken in order to stay on course with cost targets. Waiting just a few months to find out that the designs are off course can be deadly to a company because there will not be enough time cushion to adjust. In some industries, the complete product development cycle may be as short as a few months.

Since more and more design detail is available with each checkpoint, the cost estimates can be fine-tuned and become more and more accurate. The company's cost estimators must develop estimates for each purchased component, raw material, capital investment, and tool to be able to provide an even better knowledge to the product development team. The accuracy must now narrow down to within 5% or better to enable an effective, well-informed decision-making process. This might be difficult when the designs are still not finalized, but having a well-modeled estimate of cost is infinitely better than having no estimate at all. Otherwise, the engineers will be "flying blind" and will proceed with product performance as their only guiding requirement.

SUPPLIER AND IN-HOUSE MANUFACTURING QUOTE EVALUATION

As the company gets deeper into the product development process, the designs become mature enough to get feedback on cost from suppliers and the company's own manufacturing engineers. If a company has done a good job up front in evaluating designs for cost, then there should not be any surprises from suppliers and in-house manufacturing. Very often, however, companies find out only at this stage that their designs are over target. For those companies, this is the first time they are able to get any feedback on cost. Without any cost estimating capability, a company will have to rely on suppliers to tell them what the costs will be, and this is already after the designs are complete.

Even in cases where a company followed the cost evaluation process from concept, the supplier and in-house manufacturing feedback might prove that the cost is completely off target. This is usually due to the fact that initial reaction by suppliers and in-house manufacturing is to be very conservative in their cost estimates. This phenomena is sometimes referred to as "sand bagging"; assuming extra cost just in case something goes wrong. That is especially true when the designs are new to the company and its supply base, because there is a perceived risk in making the components for the first time without the designs being completely validated. It also occurs when a company does not have a trusted relationship with its suppliers. Suppliers might perceive a risk of working with a particular company based on previous experiences, often because they had to absorb some cost in the past due to their customer's negligence and/or demands. The same could be said of a company's own in-house manufacturing plants and factories. With any new design, the manufacturing engineering team will probably be conservative with the required investment and manpower in order to protect itself against any mistakes in assumptions.

This is why it is critical to also have the company's cost estimators evaluate quotes at this stage. The estimators should develop their own estimates of the designs, then compare those estimates against the estimates provided by suppliers and in-house manufacturing plants. Assuming that both provided good cost breakdowns (e.g., details of cost for raw material, labor, overhead, etc.), the cost estimators should be able to analyze the differences and identify what is driving the gaps. After gap drivers are identified, the team can work to eliminate them through fine tuning, negotiation, or by updating cost estimates to reflect any missed or incorrect assumptions.

In the SPARK hybrid truck example, where the company received quotes from suppliers and its manufacturing plant that put the vehicle $40,000 over its $90,000 cost target, what should ensue is the comparison of those quotes to internal cost estimates. With the cost estimates as an evaluation tool, it would have understood what portion of that $40,000 is due to over design and what portion is due to suppliers/in-house manufacturing plants "sand bagging." For example, if the cost estimators came up with a "should be" cost of $115,000 for the vehicle, it is likely that the excess cost in the product is caused by a combination of over design by roughly $25,000, while the remaining $15,000 is due to "sand bagging." Without quality cost engineering/estimating information available, the company will have no idea what is driving the $40,000 in cost overrun.

Not only should a company have a cost estimating team evaluate cost at each product development checkpoint, it should also build in formal checkpoints into its product development process where the supplier/in-house manufacturing feedback is evaluated. Since suppliers and in-house plants are critical partners in product development, they should be given as much time as possible to fine-tune their designs, manufacturing processes, and costs. They should not be left in the dark until the designs are finalized.

PRODUCT LAUNCH AND PRODUCTION

As the product launches into production and its design is being fine-tuned, it is still important to evaluate and control cost at each possible stage. At this stage, most of the cost is already designed in, but there is still opportunity for cost-reducing changes in design and/or manufacturing process either in-house or at the suppliers. This is an especially important stage in the cost control process because late changes made to the design in order to meet performance challenges are often key drivers of premium costs from suppliers. With tight timing and all the leverage, suppliers will often price changes at very conservative levels. In addition, since most purchasing departments are judged on annual savings achieved and not on design change pricing, buyers will have little incentive to negotiate these price increases. In fact, any such premium pricing will only help buyers once the part goes into production because it provides a larger supplier profit margin to negotiate on for future annual savings.

It should be apparent that it is critical to employ cost estimators during this stage as well. Having accurate cost estimates available for design changes will enable better control of the cost during the change management process by alerting the product development team to higher than expected cost increases. This will help prevent cost overruns as the product launches into production.

VALUE ENGINEERING

As you can see in Figure 7.1, what happens between each product development checkpoint should be a value engineering effort (cost optimization). With the cost being evaluated at each checkpoint, starting with the initial concept, the company should have a good idea of where the cost is versus target. If the cost is above target, engineering will need help in identifying cost reduction ideas. As was stated earlier, the value engineering methodology, using functional analysis principles, benchmarking idea brainstorming sessions, or TRIZ, can be a powerful tool for engineering.

VALUE ANALYSIS

The methodology of value analysis is very similar to value engineering, but is meant for products that are already in production, as noted on Figure 7.1. Nevertheless, the tools used for cost reduction idea generation in value analysis are identical to those used in value engineering.

As was mentioned before, focusing only on value analysis in the product development cycle is a terrible idea. It is very difficult to make design changes after the product goes into production; all the tools have been fabricated and any changes will require expensive tooling modifications or, for major changes, completely new tools. In some cases, the savings will need to be shared with suppliers in order to motivate them. Finally, for those companies selling to original equipment manufacturers (OEMs), they will also require their customer's approval and most likely will need to share some of the savings. As a consequence, the ROI is often not good enough to pursue changes after production start.

It is important to challenge yourself as a company to look for cost reduction opportunities throughout the whole product life cycle. It is often true that cost reduction opportunities identified during the product development process cannot be implemented due to timing, thereby making implementation during production the only possibility. Other cost reduction ideas might still have a very good ROI even if the savings must be shared with a supplier or a customer. There are also cost reduction ideas that might not be implementable during production but can be used in the next product generation. Therefore, the effort to find cost reduction opportunities is rarely wasted even if the product is already in production.

SPARK EXAMPLE

Let's go through the proposed process step by step using our SPARK hybrid truck as a case study.

CONCEPT DESIGN

As the first step, the finance team develops target costs for the vehicle, including targets for each individual vehicle component. Then, after engineering comes up with the concept design, cost estimates are developed (taking low volume of this product into consideration) for that design and compared to target costs per Table 7.1. As was already known from Chapter 1, the cost of initial concept for the hybrid truck is way over the target, which is all driven by the cost of purchased components. This design would result in a loss of $15,000 per vehicle.

Now, instead of proceeding with the concept as is, the designers have a chance to re-evaluate the BOM and get it closer to the cost target of $75,000. The designers now realize that due to the presence of a gasoline engine, the battery is difficult to fit into the vehicle and drives its shape to be difficult to manufacture, which increases its cost. Also, the extra weight of both powertrains drive the frame to be much more robust that initially anticipated. With this information, the company decides that a purely electric vehicle is the only feasible option that could meet the $100,000 market price requirement. This is a drastic change in design direction, but proceeding with the original hybrid design would have been disastrous for the company.

TABLE 7.1
SPARK Cost Status at Concept Design Checkpoint

	Target Cost	Concept Estimated	Gap to Target
Motor 1	$1,500	$2,500	$1,000
Motor 1	$1,500	$2,500	$1,000
Inverter 1	$1,000	$1,500	$500
Inverter 2	$1,000	$1,500	$500
Battery	$5,000	$13,000	$8,000
Gasoline engine	$10,000	$10,000	$0
Electronic controls	$2,000	$3,000	$1,000
Frame	$15,000	$23,000	$8,000
Body	$10,000	$15,000	$5,000
Lighting	$1,500	$1,500	$0
Other	$26,500	$26,500	$0
Total purchased parts	**$75,000**	**$100,000**	**$25,000**
Labor (factory)	$2,000	$2,000	$0
Overhead (factory)	$8,000	$8,000	$0
Other	$1,000	$1,000	$0
SG&A	$4,000	$4,000	$0
Profit	$10,000	($15,000)	($25,000)
Target price	**$100,000**	**$100,000**	**$0**

CONCEPT FINALIZATION

After the engineering team goes back to the drawing board, they develop a fully electric vehicle concept design. Per Table 7.2, this concept still does not meet the $10,000 profit requirement because the BOM cost is $9,000 above its target, but the company now has a feasible starting point to work with in order to make this vehicle a viable business.

Prior to finalizing the design, engineering decides to hold a cost reduction workshop with other company functions in order to further reduce the BOM cost. It hosts a 5-day workshop that utilizes functional analysis, TRIZ, and some benchmarking data. Based on this workshop, the cross-functional team comes up with 350 cost reduction ideas, with 210 of those ideas deemed feasible and 75 of them deemed a priority for further investigation. After analysis, Engineering decides to implement 35 of the ideas mostly related to the electronic connection concepts between battery, motors, and their electronic controllers. Table 7.3 shows an updated BOM cost analysis.

This final concept design is now projected to have a $5,000 profit, which is a significant improvement, but it's still only half of the profit target. The company, nevertheless, decides to proceed to the next phase of the project with a direction to Engineering to find further improvements in the design as drawings are developed.

In addition, since SPARK follows the collaborative product development process, the relationship with suppliers is very good and suppliers are now engaged in the

TABLE 7.2
SPARK Cost Status at Concept Design Checkpoint (Updated)

	Target Cost	Concept Estimated	Gap to Target
Motor 1	$1,500	$2,000	$500
Motor 1	$1,500	$2,000	$500
Inverter 1	$1,000	$1,250	$250
Inverter 2	$1,000	$1,250	$250
Battery	$15,000	$20,000	$5,000
Gasoline engine	$0	$0	$0
Electronic controls	$2,000	$2,500	$500
Frame	$15,000	$16,000	$1,000
Body	$10,000	$11,000	$1,000
Lighting	$1,500	$1,500	$0
Other	$26,500	$26,500	$0
Total purchased parts	**$75,000**	**$84,000**	**$9,000**
Labor (factory)	$2,000	$2,000	$0
Overhead (factory)	$8,000	$8,000	$0
Other	$1,000	$1,000	$0
SG&A	$4,000	$4,000	$0
Profit	$10,000	$1,000	($9,000)
Target price	**$100,000**	**$100,000**	**$0**

TABLE 7.3
SPARK Cost Status at Concept Finalization Checkpoint

	Target Cost	Concept Estimated	Gap to Target
Motor 1	$1,500	$2,000	$500
Motor 1	$1,500	$2,000	$500
Inverter 1	$1,000	$1,000	$0
Inverter 2	$1,000	$1,000	$0
Battery	$15,000	$18,000	$3,000
Gasoline engine	$0	$0	$0
Electronic controls	$2,000	$1,500	($500)
Frame	$15,000	$16,000	$1,000
Body	$10,000	$11,000	$1,000
Lighting	$1,500	$1,500	$0
Other	$26,500	$26,000	($500)
Total purchased parts	**$75,000**	**$80,000**	**$5,000**
Labor (factory)	$2,000	$2,000	$0
Overhead (factory)	$8,000	$8,000	$0
Other	$1,000	$1,000	$0
SG&A	$4,000	$4,000	$0
Profit	$10,000	$5,000	($5,000)
Target price	**$100,000**	**$100,000**	**$0**

development process. In fact, they could have been involved directly during the early concept design stage. The best way to do this is to bring in the experts from suppliers for a cost optimization workshop, which expands the brainpower pool to more individuals and ones that are experts in their respective products and manufacturing processes. For example, a supplier producing a stamping for the final assembly is an expert at stamping and should know more about alternative materials and/or manufacturing processes. This kind of expertise is critical in fine tuning the design and the cost estimate of a concept and, more importantly, finding cheaper ways to meet required functionality.

The suppliers were not brought in for the initial cost reduction workshop but are now asked to provide their feedback on the design, and while Engineering is working on developing drawings for their initial design, the feedback from suppliers comes in to support that development.

Phase 1 Design (Quote Level)

After receiving feedback from suppliers, the designers now make final design adjustments and release the official 2D and 3D drawings. Those are then sent out to suppliers as a request for quotation (RFQ) package. Table 7.4 is a summary of gaps after the quote response from suppliers and SPARK's manufacturing plant is received and an update to the cost target estimate is made. The good news is that the estimates

TABLE 7.4

SPARK Cost Status at Phase 1 Design Checkpoint

	Target Cost	Phase 1 Estimated	Phase 1 Quote	Gap to Target	Gap to Quote
Motor 1	$1,500	$1,756	$1,955	$256	$455
Motor 1	$1,500	$1,756	$1,955	$256	$455
Inverter 1	$1,000	$951	$1,125	($49)	$125
Inverter 2	$1,000	$951	$1,125	($49)	$125
Battery	$15,000	$16,597	$17,567	$1,597	$2,567
Gasoline engine	$0	$0	$0	$0	$0
Electronic controls	$2,000	$1,850	$2,156	($150)	$156
Frame	$15,000	$15,098	$17,231	$98	$2,231
Body	$10,000	$10,250	$11,422	$250	$1,422
Lighting	$1,500	$1,300	$1,328	($200)	($172)
Other	$26,500	$25,124	$27,894	($1,376)	$1,394
Total purchased parts	**$75,000**	**$75,633**	**$83,758**	**$633**	**$8,758**
Labor (factory)	$2,000	$2,000	$2,145	$0	$145
Overhead (factory)	$8,000	$8,000	$8,745	$0	$745
Other	$1,000	$1,200	$1,244	$200	$244
SG&A	$4,000	$4,000	$4,000	$0	$0
Profit	$10,000	$9,167	$108	($833)	($9,892)
Target price	**$100,000**	**$100,000**	**$100,000**	**$0**	**$0**

confirm an improved profit of $9,167 is possible, but the bad news is that the quotes bring the profit down to only $108, which is not acceptable.

Typically what happens in this situation at most companies is that Engineering would "wash their hands" off of it and put it on Purchasing to negotiate suppliers down or have other functions "sharpen their pencils." Engineering would claim that they have done the best they could and there is no possibility to improve the design any further. Although both Purchasing and other functions will probably do the best that they can, especially since all these estimates are still just estimates and could easily be off by 10% or more each way, Engineering should continue to be engaged in the cost optimization process. It is very common, even at this stage, after so much analysis has taken place, that the opportunities to further improve the design still exist and may be significant. In addition to regular design and manufacturability reviews, it is critical to hold further cost optimization workshops, even if it means putting stress on the product development timing or adding more resources.

The company decides to hold another workshop, this time together with suppliers to find additional cost reduction opportunities. Since the functional analysis was already performed in the prior workshop, this new workshop reviews the results of the prior workshop and focuses more on the value stream map for this product and any lean manufacturing opportunities.

At the same time, Purchasing and Cost Estimating engage suppliers in a detailed analysis of their estimated pricing. Gap analysis is performed comparing quotes to cost targets that evaluates every cost bucket assumption for every vehicle part. This identifies every gap driver and establishes action plans to close those gaps. The action plan items could be opportunities in both design and manufacturing process. These are then evaluated with Engineering for updates to the design. The same gap analysis process is done with SPARK's own factory since their costs came in higher than expected. In addition, lean manufacturing study is performed to optimize the process.

It is important to note that a certain level of conservatism is expected in first time quotes since both suppliers and SPARK's factory face high levels of uncertainty and risk surrounding a preliminary design, especially for something as revolutionary as electric vehicle components. This is why a collaborative problem solving approach using gap analysis is key to help understand the assumptions and to reduce the amount of risk assumed.

FROM PHASE 2 TO FINAL DESIGN

Using the results of the workshop plus the results of gap analysis with suppliers and SPARK's factory, Engineering now continues to fine tune the design repeating this exercise through each design iteration until the very last one (see results in Table 7.5). The company is satisfied with achieving $8,693 in projected profit, even though it's slightly below target and with possible estimating inaccuracies. It hopes to achieve

TABLE 7.5
SPARK Cost Status at Final Design Checkpoint

	Target Cost	Final Estimated	Final Quote	Gap to Target	Gap to Quote
Motor 1	$1,500	$1,756	$1,825	$256	$325
Motor 1	$1,500	$1,756	$1,825	$256	$325
Inverter 1	$1,000	$894	$900	($106)	($100)
Inverter 2	$1,000	$894	$900	($106)	($100)
Battery	$15,000	$16,258	$16,667	$1,258	$1,667
Gasoline engine	$0	$0	$0	$0	$0
Electronic controls	$2,000	$1,750	$1,923	($250)	($77)
Frame	$15,000	$14,963	$15,247	($37)	$247
Body	$10,000	$10,250	$10,536	$250	$536
Lighting	$1,500	$1,300	$1,328	($200)	($172)
Other	$26,500	$24,620	$24,954	($1,880)	($1,546)
Total purchased parts	**$75,000**	**$74,441**	**$76,105**	**($559)**	**$1,105**
Labor (factory)	$2,000	$2,000	$1,958	$0	($42)
Overhead (factory)	$8,000	$8,000	$8,000	$0	$0
Other	$1,000	$1,200	$1,244	$200	$244
SG&A	$4,000	$4,000	$4,000	$0	$0
Profit	$10,000	$10,359	$8,693	$359	($1,307)
Target price	**$100,000**	**$100,000**	**$100,000**	**$0**	**$0**

an even higher profit as it launches the product, especially considering that the cost targets indicate that $10,359 in profit is achievable.

Having confidence in profit achievement is critical at this stage since production tooling will now be kicked off. Any further design changes would cause significant investment in new or updated tools and equipment, which in turn would diminish the ROI. Changes will also cause havoc on supply chains and manufacturing plants since it can be difficult to keep track of various parts on order and can ultimately lead to quality issues. Finally, as a supplier looking for improvements after tools are built and design validation completed, any changes will require customer approvals and perhaps share of the savings as an incentive for approvals. Customers are not keen on making changes after a certain point in product development process because changes pose significant quality and delivery risks which can delay program timing. Also, even though companies are usually unable to measure it, there are significant "hidden" costs outside of tools and equipment, such as resources needed to execute changes.

As the design changes to address product performance issues, the development team should constantly evaluate product cost versus target and continue to make adjustments as needed. In some cases, where product design or manufacturing process changes need to be significant, another cost optimization workshop with cross-functional team and suppliers might be required. It is even advisable to sometimes bring in outside consultants, or company engineers that have not been involved already, to obtain a fresh perspective about the design.

Production Phase

If a design is optimized to meet or beat the cost target, it is now safe to kick off the production tools. After the parts pass all the functionality and durability tests, a profitable product can be launched into production. Although no further cost optimization activity is needed for this product to meet profitability requirements, it is critical that the activity continues after the start of production. Typically, in a life cycle of a product, there is limited time between production releases of different product generations. For example, in the case of automobiles, new versions of the same car are released every 4 or 5 years; however, there are usually vehicle refresh updates every 2 to 3 years. Recently, with the advancements in CAD (computer-aided design) and CAM (computer-aided manufacturing), the release times for these car updates are shrinking even further. This means that automotive OEMs must find ways to improve value and cost within a shorter period of time. In other industries, like appliances and consumer electronics, the product life cycles can be even shorter and a product might have to be refreshed every few months.

The product designers, therefore, must use the time between production releases wisely. Instead of waiting for a certain period of time to start working on a new design version, the designers must start developing cost effective refresh concepts as soon as production starts. With the product launch still fresh in everyone's mind, it would be wise to collect lessons learned to implement in the future. As part of that effort, a valuable tool at this time might be benchmarking, which gives a valuable comparison on how a recently launched product stacks up against products recently launched by the competition.

Since designers are so intimate with their designs, there sometimes might be a resentment for having to consider somebody else's designs, but it is important to keep in mind that the number one goal of any company is to make money, not to make egos. Developing designs is only a means to an end, the profit goal must always be a priority. The pride in a design must be tossed aside and replaced with the pride in having the company make money and everyone keeping well paid jobs.

The results of benchmarking and other cost optimization activities are then taken into consideration for the next product iteration. As the designers start their work on the next concept, they now have a list of optimization ideas to consider. Although benchmarking usually involves only current competitor products, and not their future designs, it is still important to consider new creative solutions in the concept development. Using benchmarking data assures that your products will have the most time to catch up to competition.

REAL-LIFE ANECDOTE

A Tier 1 supplier in the automotive industry was tasked by its executive management to increase the amount of savings on its current production products, but since the existing organization and processes were under-delivering, this meant changing how things were done. Unfortunately, there were many issues in implementing changes. First, the CEO was not engaged on this activity, which left the different functional VPs to bicker over how to implement it. It also left the working level teams with a lack of clear direction. In fact, the engineering function went as far as sabotaging this effort so that its designs were not questioned and its resources not distracted from its main design responsibility. Second, no dedicated resources were established to execute the changes. The small organization that was established consisted of people that already had other primary jobs. Finally, the objectives of the functions and participants were not aligned to reflect the cost optimization goals. Most of the functional objectives remained the same, which left functions fighting over credit for each cost reduction. The result of these issues was that ultimately nothing changed and the organization kept underperforming on cost savings for current production parts.

SUMMARY

It is critical to employ cost engineering in the product development process from concept to production and back to concept. It is a form of cost control, giving product development teams both knowledge of their product's cost as well as tools to do something about it. As much as engineers love to have a free hand in the creative process and an ability to develop awesome products, they also love to keep their jobs, which is difficult to accomplish when the products are not making any money for the company.

There are a lot of skeptics that will argue that they do not need the cost engineering resources and processes at their companies because their products don't change and historical data is sufficient to guide their engineers. This is, however, seldom the case. In reality, products constantly evolve. In order to keep up with market trends and competition, companies are forced to continuously change their products.

For example, cars with four wheels and gasoline engines have been produced for over a hundred years, yet the technology for each and every vehicle component has evolved dramatically. Engines are constantly improving to get more torque using less gasoline, seats are now movable in almost all directions and have heating and sometimes cooling, the electronic controls changed from push buttons to touch screens, car bodies are moving away from steel into aluminum and plastic, and now powertrains are shifting from gasoline to hybrid or fully electric.

The same can be said of almost every industry – consumer electronics, aerospace, defense, shipping, food, even toilet paper – all are going through constant change and some even through very dramatic transformations.

With the continual advances in electrification and automation, it is more likely that the pace of product change will speed up going forward. Companies will have to do more development in less time to keep up. Although it is an additional step in the product development process, cost engineering capability will become even more critical to the well-being of any company. It is really the only way to avoid catastrophic cost overruns and sustain a company's success in the long run.

8 Cost Engineering Organization

Cost engineering efforts at most companies can only be described, at best, as sporadic, inconsistent, and mostly ineffective. Cost engineering resources are sometimes placed in engineering, purchasing, finance, or operations, but rarely branch out to all those functions simultaneously and rarely encompass the complete life cycle of a product. The intensity and scope of the cost engineering effort is usually very person-dependent. This is where a leader of one of the functions, perhaps with previous cost engineering experience, might hire one or two people and charge them with cost reduction activities.

For example, a head of engineering might be tasked with improving the cost of a company's designs, so he will add a person with cost optimization experience to his staff. That person will then have to lead all cost optimization initiatives. Unfortunately, he/she will usually not have the organization's cooperation since his/her work is out of scope of the engineering process. He/she will not have any real authority in engineering or outside of it and will have to work within established processes and procedures that are usually not designed to support cost optimization. In addition, nobody in the organization will have objectives aligned with cost optimization, so engineers will usually continue to focus only on their main objective of designing a product that works. The probability of successfully engaging the entire organization in cost optimization efforts are therefore very limited to none at all.

The lone cost optimization engineer might find some opportunities and might even be able to convince others to implement them, but it is rarely a sustainable effort. In some cases, when great success was achieved in implementing savings, an organization might hire a few more cost optimization engineers. This small group of people will continue to look for opportunities, usually on existing products, and have some success in implementing their ideas. That is, until the head of engineering leaves his position for another opportunity. The new head of engineering is hired and he/she decides that more resources are needed in core design, so the cost optimization engineers are transferred (if not fired) to the design group. That is usually the end of the company's cost optimization efforts.

Whether the above scenario takes place in engineering, or purchasing, or finance, or operations, the result is usually the same, an unsustainable, isolated, and usually unsuccessful cost optimization effort. As long as cost optimization is person-dependent and considered only as an optional non-core capability, this will never be a successful endeavor.

The good news is that there seems to be a recent trend in improving the state of cost engineering and optimization at companies around the world. As the economy becomes more and more global, the competition is intensifying and leading to increased focus on cost. Some Japanese firms have already been held in high regard for their continuous improvement culture and cost planning techniques (such as target costing covered earlier in the book). These methodologies are being used by an increasing number of companies around the world. Their main advantage is to have cost discipline throughout the product development process. This helps companies avoid the deadly cost overruns late in the development process. When those overruns do occur, target costing methodology forces companies to spend time and effort on optimizing products so they reach production on budget. This kind of discipline is derived by having a dedicated cost planning group within the finance organization that might report directly to the CFO. This way, the importance of the cost optimization discipline is clearly established and the authority and objectives are aligned with other functions.

While using the target costing philosophies, some companies are also taking cost optimization to the next level. Their cost engineering organization reports to the CFO, but is expanded outside of finance and encompasses resources in purchasing, engineering, finance, and operations. These might report in a matrix structure to both the CFO and to the heads of those functions and/or to the heads of various product lines (see Figure 8.1). With this organizational structure, the whole company is engaged in the cost optimization effort.

For example, a company would have cost optimization experts embedded in the following functions performing specific responsibilities:

Engineering – working with designers and engineers to optimize designs for cost, especially early in the development process

Purchasing – working with suppliers to achieve optimal cost including logistics

Operations – working with in-house manufacturing engineers on achieving optimal processing cost

Finance – bringing costing discipline and control to other functions and product lines as well as consolidating cost data to develop business cases and quotes to customers

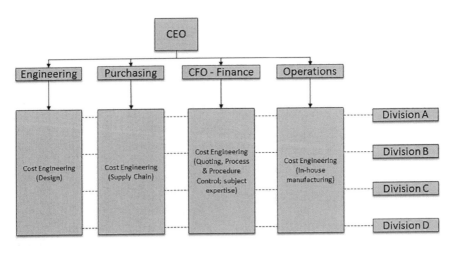

FIGURE 8.1   Cost engineering organization.

All of these cost optimization resources will report to their respective functions or product lines, and also to the head of cost engineering who in turn will report directly to the CFO. Why report to the CFO and not the other functional heads? This is because the cost engineering effort is considered a cost controlling function, the same way an accounting controller would be charged with cost and profit reporting discipline of the organization. As a cost controlling function, cost engineering is ultimately tied to the customer quoting function (this is sometimes called business controlling), also under Finance, which helps drive the most competitive customer quotes.

It is correct to assume that most of the cost optimization work takes place in various functions and product lines/business units; however, the head of cost engineering will have a staff of people who will be responsible for implementing standards for processes and procedures. His/her staff will also include experts in various cost estimating and optimization techniques that can be deployed to various areas of the company to support projects.

As efficient and successful as these benchmark organizations might be, it does not mean that there are no better ways to structure cost engineering. As long as the role of cost engineering is respected and part of the company core culture, the organizational evolution can continue. Perhaps the head of cost engineering should report directly to the CEO. Since cost management is such a critical function, just as much as engineering and purchasing, why not give it an equal status?

Some might say that reporting under Finance is a perfect fit because it is somewhat similar to cost accounting, but accounting usually focuses on the past or short-term future while cost optimization and estimating focus more on mid- to long-term future. Others would argue that reporting under Purchasing or Engineering is ideal; however, the risk of being pulled into their core activities could be too much to outweigh any advantages of having proximity to those functions.

COST ENGINEERING CULTURE

The importance of cost engineering culture was mentioned as one of the keys to success, but this is often forgotten or misunderstood. It is true that most companies struggle with establishing their preferred cultures just in general, so it is not surprising that establishing cost engineering culture would not be any different. Without direction, cultures normally develop on their own, like an ecosystem, based on the people that work in a given company. Culture is not a poster in a conference room that states what the management wants the culture to be; it is what actually happens in the conference room. It is, therefore, important for management to provide cultural direction and then drive that culture on a daily basis.

In the case of cost engineering, the management team, starting with the CEO, must stress the importance of cost engineering. They must actually build it into the product development process and follow the process by insisting that all the steps be completed correctly. A company should include cost engineering reviews at each critical step of its product development process starting with concept design and ending with product launch. At each step, or gate, the cross-functional team must review cost status versus target and develop a plan to close the gap to target using cost optimization tools mentioned in previous chapter. A list of cost reduction ideas must be developed together with a plan of implementation for each. Clear roles and responsibilities must be assigned.

Another part of cost engineering culture that is required for success is openness to change. This is much more difficult than it sounds, especially for larger companies. In order to control its operations, companies will implement systems and procedures on its workforce that inhibit change. The larger the company, the more restrictions it tends to impose on itself, sometimes even mimicking country governments in its bureaucracy.

These rules and regulations aim to prevent failures in product execution and allow for certifications of various quality standards that customers and industries require. They also restrict variances from the standards. This ultimately drives creativity out of the organization even though creativity is essential to innovation. It is therefore critical for the company to drive the culture of change and to embrace change and continuous improvement as one of its pillars of existence. This will not only allow for cost optimization to be embraced, but also it will serve the company well in general by making it more agile to the ever-changing business environment. A great reference on driving organizational change is a book "Switch: How to Change Things When Change Is Hard" by Chip Heath and Dan Heath.

Over time, the discipline of enforcing the cost engineering process and continuous improvement mindset will lead to culture change. The company's employees will be compelled to make it part of their mindset and will not have a choice to develop their own approach around past experiences and preferences. The management team must strive to have the discipline, but the success that cost engineering will create will ultimately lead to acceptance of the cost engineering culture by the staff.

DEVELOPING COST ENGINEERING CAPABILITY

Cost engineering talent is in short supply around the world due to increasing demand for cost engineering and optimization throughout various industries. At the same time, there is continuous lack of professional training at universities, colleges, and vocational schools. In most countries, since cost engineering has not been acknowledged as a core function, the educational institutions have followed suit and ignored it as a field of study. Most companies also do not have any training or apprenticeship programs for cost estimating or optimization experts. Therefore, cost engineering is usually self-taught or trained on the job.

Worse yet, there is usually no specific career path to get into the cost engineering field or to move up from there. It is often considered a dead-end job because those in core functions do not view cost engineers as having enough core function experience. This is despite the fact that individuals engaged in cost engineering actually acquire much more knowledge about company operations, its suppliers, products, and manufacturing processes than almost anyone else in the company.

Therefore, for a company to have a sustainable cost engineering team, it must put in place its own training and development plan. With cost engineering as a core function, it is key that a company begins to develop individuals early in their careers, perhaps as early as an internship or apprenticeship program and through a career path that perhaps leads through engineering, manufacturing, purchasing, finance, or all of the above to give this individual all the necessary knowledge that he/she will be using in the cost engineering function. Throughout this development, it will be important to provide individuals with training specific to cost engineering. Since cost engineering concepts are not part of the university curriculum, a full training course (including mentorship program) will need to be developed. This training will need to cover all the concepts and available knowledge, such as cost estimating and cost optimization techniques. It would not be a stretch to expect this training material to span an equivalent of two elective courses in college. Although such an undertaking might seem daunting for a company, it will be worth the investment as it will create highly skilled individuals that can help generate extremely cost competitive products and ultimately much more profit for the company.

It is also important to provide individuals working in cost engineering with a career advancement path. Just like those in Engineering, Purchasing, Manufacturing, or Finance, each individual in Cost Engineering must feel that he/she can advance from junior to senior cost engineer, then on to supervisor and manager, director, and perhaps a vice president or even a chief executive. Without this option, only those that are satisfied with simply remaining a cost engineer for their whole career will stay in the field. Those who want to be leaders and influencers will eventually leave cost engineering or never get into it in the first place. This is not any different than an engineer wanting to lead an engineering team, instead of being a staff engineer for 40 years.

Finally, it is important that an organization values its cost engineering professionals enough that they are also considered for positions in other functions. Just like a company would want an engineer to be given opportunities in Purchasing or Operations, a cost engineer should also be considered for jobs in other functions. Too often, a company will consider the cost engineering expertise so unique that it cannot foresee this skill set to be translatable to other functions. This is a mistake and the opposite is true. As cost engineering experts acquire their cross-functional knowledge while performing their jobs, they usually have a much wider understanding of linkage between functions and products. For example, a cost engineer that wants to be a buyer might not have the background of all the buyer's daily tasks, but his/her overall understanding of designs, manufacturing processes, and cost structures will normally far outweigh the knowledge of a typical buyer. Similarly, a cost engineering manager moving into a management role in other functions will also have a huge benefit of cross-functional and product/manufacturing knowledge, not limited to experience in just one function, regardless of how detailed that one function experience might be.

One could even argue that cost engineering professionals should be fast tracked or prioritized for promotions. Just like a CEO's development usually involves gathering working experiences in various core functions in order to provide an individual with a full understanding of how the company functions and how products are developed and produced, so does the cost engineer gather this kind of knowledge that can be used to better manage teams, groups, divisions, or product lines.

A typical cost engineer or manager will spend a lot of time at its own or suppliers' manufacturing plants where he/she will learn in detail about manufacturing processes, factory organization, and its finances. At the same time, this individual will work closely with engineering to understand designs down to detailed GD&T level, learning a lot about product development, design, testing, and launch. But that's not all. This same cost engineer or manager will also work closely with Purchasing where he/she will most likely be part of the commercial negotiations and supplier relationship building. Finally, this individual will have to work with finance and/ or accounting functions either in-house or at suppliers, which gives him/her a lot of knowledge on cost structures and cost allocations as well as general knowledge of financial reporting.

The cost engineering professional will, through daily work, gain in-depth knowledge on literally all other core functions: engineering, manufacturing, purchasing, and finance. No buyer or engineer will ever gain this kind of cross-functional knowledge working only in their functions. If knowledge diversification is key to a company's success, then why shouldn't cost engineering professionals be considered or even fast tracked for positions in other functions?

COST ENGINEER INTERVIEW

Can any person be a good cost engineer? The answer is no, because just like those who are good at marketing might not be good at engineering, those who are good at engineering might not be a good fit for cost engineering. The personality traits most applicable to cost engineering are typically unique to that function, even though the reverse might not be true.

As an interviewer in the cost engineering function, what should you be looking for in a candidate that will make him or her successful? If you asked this question to cost engineering professionals, most would probably answer that an analytical mindset and an attention to detail are most important. Although these two are indeed very important, the most critical trait for a cost engineer is CURIOSITY. Whether you are cost estimating or cost optimizing, curiosity to understand what is behind the assumptions and calculations will drive the most success in cost engineering. Those who try to understand will get closer to the most truthful and accurate answers and will also find the most optimized solutions to problems.

Another important trait, especially for those in cost optimization, is the ability to challenge the status quo and be the change agent for an organization. This trait is needed because even if an optimal solution is found, most people do not like change and will resist it, so the ability to convince and motivate people to make a change in design or manufacturing process is very valuable. A change agent must also be able to provide the right tools and show the path on how to get to the desired results.

The question is, how do you know an individual possesses the above traits? Perhaps it is easy to tell from a resume if an individual has had prior experience in cost engineering, but what if that is not the case? The answer is that if you want someone for the job who is curious then you should look in the resume or during the interview for things that showcase this personality trait, such as interests in travel and history. Or, if you are looking for someone who challenges the status quo, then you should ask if that person has ever proposed any changes and how they drove them to completion. Or, ask if the interviewee was part of an organization that seeks change in the society or the community. Are they able to speak to large groups and sell things or concepts?

As you might have noticed, although curiosity and change agent ability are preferred for cost engineers, these traits are also very beneficial in other functions of the organization. As was mentioned before, even though cost engineering skills are unique, the individual possessing them can use those skills to drive success in other functions and should even be fast tracked through their career path.

COST ENGINEERING SOFTWARE

An evolutionary component of cost engineering is coming from the software industry. Some software companies – such as FACTON, Siemens, and aPriori – have developed increasingly sophisticated software that gives companies capabilities in cost estimating, cost allocation, target costing, and business case development for customer quoting. There is also software available with the capability to estimate tooling cost, such as Schmale, Siemens, and aPriori. These software programs contain thousands of raw material, labor, and machine rates for various regions of the world, which can save companies a lot of time trying to collect this information themselves.

More recently, companies such as aPriori, Costimator, and Munro have taken their software to a higher level by putting real time estimating capability at the engineers' fingertips. These software can take 3D models and estimate part cost without any input from cost estimators or suppliers. Instead of taking days or weeks to have estimates developed by expert cost estimators, engineers can learn the impacts of their design decisions immediately.

Perhaps the next evolution will be to help engineers even more by providing design problem solving suggestions based on functional analysis or TRIZ concepts discussed earlier in this book. Instead of spending valuable time and resources on organizing brainstorming workshops, future software will automatically suggest cost optimized solutions.

SUMMARY

In order to have a long term success, a company must develop a cost engineering organization within its ranks. This organization must be embedded cross functionally so that all key functions of the company have individuals dedicated to developing cost optimized products. In addition, cost engineering must be part of the company's culture so it is not just a few individuals that are dedicated to the task. It must be a company wide effort. The building block of this should be incorporation of cost engineering checks throughout the product development process so that cost status and action plan to close gaps to target are evaluated early and often. Implementing such a culture will require effort and discipline from the management team starting at the top with the CEO.

It is important for a company to develop and foster its cost engineering capability. Finding the individuals with matching talents and then training them are key parts of that capability development, but it will also require developing and promoting the people in and out as well as within the cost engineering organization. There must be a clear career path for individuals that are interested in cost engineering and the wide ranging knowledge and skills they acquire in their cost engineering roles should be viewed as assets to any function within the organization.

9 Cost Engineering in Conclusion

The future of cost engineering is bright and already here. The companies with the most efficient cost engineering organizations and tools will be the most competitive and successful companies of the future. Cost engineering can be such a powerful competitive advantage that others will simply not be able to keep up. Some might argue that their company does not have the budget for resources above and beyond their core functions. The honest answer is that **cost engineering is a core function**, whether you acknowledge it or not, and you better find the budget for it. Any company manufacturing any product must already cost optimize it in order to make money; some are just doing it better than others. A consumer is willing to pay only so much for any given product, which means that a cost target is automatically established. Developing a product that costs more than the target means eroding or completely wiping out profit, therefore the product must be cost optimized to the target or better. If your company is not doing this basic thing, then it should not and will not be in business.

If it is established that cost engineering is a core function, then it follows that the choice is only how good a company wants to be at cost engineering and how it plans to manage the resources and budget to get there. Regardless of whether a company has a dedicated team or gives software and tools to its engineers and buyers, as long as cost optimization is an organizational mindset, the chances for its success improve dramatically.

Cost engineering is not a well-known methodology and is mostly misunderstood and underappreciated, but its effectiveness is undeniable. Although there is a lack of statistics to prove it, companies that are using it will almost unanimously agree that they have benefited greatly from its use. Despite adding more complexity to the product development process, which usually means more work, the witnesses of this methodology will tell you that the rewards far outweigh the effort. The benefits are also very tangible and come in the form of profit and cash. Companies that embrace cost engineering will sell more profitable products and make their businesses more viable in the long term. They will avoid developing products that cost more than what they are sold for to their customers. They will not find themselves redesigning products mid-cycle and exceeding their development budgets. They will not end up with suppliers that are unable to manufacture components or manufacture them at a loss.

The companies that embrace cost engineering will have accurate cost data that allows them to make good decisions. They will focus on meeting cost targets throughout the product development process and work with their suppliers early to design manufacturable and cost-effective products that sell with profit on a consistent and sustainable basis.

In order for the cost engineering process to be successful and sustainable, companywide resources must be deployed during the product development process. Both cost estimating and cost optimization experts must be employed. The first to provide cost knowledge and guidance, the second to scientifically and creatively find cost reduction opportunities. These resources must span all the cost categories of the product, including raw material, purchased components from suppliers, logistics, engineering design, in-house manufacturing, and corporate overhead.

The cost engineering process must start as early as possible in the product development process, even as early as napkin sketches, and the cost engineering discipline must be maintained throughout. There must be a review of cost engineering efforts at every product development gate. The design direction must be adjusted if the cost targets are not being met and cost engineering resources must be engaged to find cost-effective solutions.

The cost engineering methodology must be fully embraced and supported. Not only does a company need to use cost engineering methodology, it must also build an effective cost engineering organization and develop people to work in it and give them career advancement options. A company must also provide the cost engineering organization with resources such as software, benchmarking facilities/ personnel, and a travel budget to visit supplier factories. Cost engineering training will have to be developed to capture and share knowledge.

It is critical to note that having cost engineering capability is often required in today's competitive world. Some of your competitors have probably already embraced it or are embracing it. Your company must do the same simply to keep up. By fully embracing it yourself, it will give you an unprecedented competitive advantage. If your company can make cost engineering part of its culture, it will stay ahead of the curve. Although cost engineering resources are not insignificant, the payoff far outweighs this investment. It is not an exaggeration to say that for every dollar spent on cost engineering, ten dollars should be expected in return. For example, a $10B company will probably need to spend $10M a year on cost engineering resources (e.g., 70 personnel, costing software, travel to supplier factories, teardown and benchmarking resources, cost reduction workshops, etc.), but should see a $100M or more improvement in profit per year, which is a great ROI.

However, companies should not get caught up in measuring this effect. If a company employs cost engineering resources early in the development process, it will find it difficult to quantify the effect that cost engineering had on the company profit. This is because the biggest effect will be in cost avoidance, not in the annual cost savings. The best way to measure cost engineering effectiveness is by quantifying the improved profit results two or three years after cost engineering implementation. If the profits for that $10B company jump from $500M annually to $600M or more in that span while everything else remains equal, then most likely cost engineering was the culprit of this dramatic improvement.

Various company functions often fight to give themselves credit for any success, but as a CEO or general manager, you must see the bigger picture. You must give credit to the cross-functional embrace of cost engineering methodology. And you must engrain cost engineering into your company culture so much that it becomes just like breathing. It must become something that your company does naturally and cannot exist without. This will give your company a method to sustain its existence even after the people who implemented cost engineering are long gone. Good luck!

SUMMARY

The future of Cost Engineering is bright and already here. The companies with the most efficient Cost Engineering organizations and tools will be the most competitive and successful companies of the future. Cost engineering can be such a powerful competitive advantage that others will simply not be able to keep up. Some might argue that their company does not have the budget for resources above and beyond their core functions. The honest answer is that **cost engineering is a core function**, whether you acknowledge it or not, and you better find the budget for it. Any company manufacturing any product must already cost optimize it in order to make money, some are just doing it better than others. A consumer is willing to pay only so much for any given product, which means that a cost target is automatically established. Developing a product that costs more than the target means eroding or completely wiping out profit, therefore the product must be cost optimized to the target or better. If your company is not doing this basic thing, then it should not and will not be in business.

Glossary

annual production hours: the total number of hours in a year that machines are able to operate; variable of shift pattern/model

bottleneck or gate station: station in the manufacturing process of a part that takes the longest time; parts cannot be produced any faster than that station; usually this station determines the cycle time of a part

burden: the overhead cost

capacity utilization: the percentage of a manufacturing facility or its equipment that is utilized for production out of the total available production time

clean sheet estimate: also referred to as "ground up" estimate, this type of estimate assumes that every cost component is estimated without using any prior historical cost assumptions as if starting with a clean sheet of paper

cost center: the collection of resources that accumulates cost such as a factory, a department, or an equipment center; objectives are usually centered on improving cost

cycle time: the time between manufacture of each part, or how fast parts come out at the end of its manufacturing process, sometimes called cycle time from "drop to drop" on the conveyor belt or collection bucket; variable of throughput

efficiency factor: the percentage of time out of total available production time that machines are producing parts

engineering, development, & testing (ED&T): this is customer-specific cost of R&D where engineering resources are used to develop existing products for specific applications only

expendable packaging: the type of packaging that is used only once to transport parts, after which it is thrown away or recycled, e.g., cardboard boxes

fixed cost: the cost that does not vary in the short to medium term, such as cost of company's buildings or its executive management staff

fringe benefits: benefits afforded company's workers, such as health and life insurance, vacations, bonuses, cafeteria and transportation services, and such

Grenzplankostenrechnung (GPK): the German costing methodology designed to provide a consistent and accurate application of how managerial costs are calculated and assigned to a product or service

inefficiency factor: the percentage of time out of total available production time that machines are not producing parts

just-in-time (JIT): Japanese manufacturing system in which the components are delivered only when they are needed for production

kanban: Japanese manufacturing system in which the supply of components is regulated through the use of an instruction card sent along the production line; the system is meant to improve efficiency and reduce inventory

long-term agreement (LTA): the agreement that a customer might sign with a supplier that secures future business with that supplier for a length of time for exchange of future price reductions or upfront lump sum payment or both

manufacturing lot: the batch of parts produced in one manufacturing run; total number of manufacturing lots per year defines number of setups that will be needed per year

margin: the difference between price and cost of goods, e.g., $2 margin on $12 price is 16.67%

markup: the amount added to the cost of goods to cover overhead and profit, e.g., 20% markup on $10 of cost is $2 ($10 cost × 1.2 = $12 price; $12 price − $10 cost = $2 markup)

OEE: overall equipment effectiveness defines the percentage of time in which good parts are produced out of the total available production time

payment terms: the terms defining how fast a supplier will get paid for its products, e.g., 90 days after customer reception of the purchased parts; higher payment terms increase part's piece price because suppliers have higher borrowing cost to account for withheld payment

planned downtime: the time that machines are down due to scheduled, or planned, breaks in production, such as lunches or machine maintenance

processing time: the time it takes to process each part from the beginning to end of the manufacturing process, e.g., 10 minutes for a part to go through an assembly process

profit center: the collection of resources that accumulates profit such as a factory, department, or equipment center; objectives are usually centered on improving profit

returnable packaging: the type of packaging that can be reused multiple times to transport parts, e.g., plastic trays

rework: the cost associated with repairing production parts that were initially scrapped; on the other hand, the cost of scrap decreases in this case

run@rate: the assessment often performed in the automotive industry to validate the supplier's ability to produce enough parts for production; usually takes place in advance of start of production

shift pattern/model: the set of production shifts and their length and timing throughout a production week, e.g., 3 shifts, 5 days a week

skilled labor: the operators who possess advanced machine operating training, e.g., machinists who operate advanced machining centers

startup cost: the cost required to launch production of a product, e.g., hiring and training of production crews prior to start of production or financing of equipment purchased in advance of production start

TAKT time: the time between manufacture of each part that needs to be achieved in order to meet the customer demand; often confused with cycle time, which could be lower in some cases

throughput: amount of parts that can be produced in any given time, e.g., 300 parts per hour

total cost of ownership (TCO): the compilation of all costs to a buyer required by a product over its total life, including concept development cost, manufacturing cost, quality and warranty costs, disposal cost, servicing cost, etc.

unplanned downtime: the time that machines are down due to abrupt or unplanned breaks in production, such as machine breakdowns or power outages

unskilled labor: the operators who do not require any advanced equipment training to operate machines

variable cost: the cost that varies in the short term with changes in production volumes, such as machine electricity or supplies

yield: percentage of good parts out of the total produced

Recommended Literature

Altshuller, G. *And Suddenly the Inventor Appeared.*

Ansari, S.L. and Bell, J.E. *Target Costing: The Next Frontier in Strategic Cost Management.*

Goldratt, E.M. *The Goal: A Process of Ongoing Improvement.*

Heath, C. and Heath, D. *Switch: How to Change Things When Change Is Hard.*

Hicks, D.T. *Activity Based Costing: Making It Work for Small and Mid-Sized Companies.*

King, P.L. and King, J.S. *Value Stream Mapping for the Process Industries: Creating a Roadmap for Lean Transformation.*

Lembersky, M. (Ed.). *Realistic Cost Estimating for Manufacturing.*

Liker, J. *The Toyota Way: 14 Management Principles from the World's Greatest Manufacturer.*

Martin, K. and Osterling, M. *Value Stream Mapping: How to Visualize Work and Align Leadership for Organizational Transformation.*

Miles, L.D. *Techniques of Value Analysis and Engineering.*

Monden, Y. *Target Costing and Kaizen Costing.*

Womack, J.P. and Jones, D.T. *Lean Thinking: Banish Waste and Create Wealth in Your Corporation.*

Index

Printed and bound by CPI Group (UK) Ltd, Croydon, CR0 4YY

24/10/2024

01778493-0007